Facility Management Practices

Solomon Pelumi Akinbogun ·
Clinton Ohis Aigbavboa ·
Oluwaseun Tope Akinbogun
Editors

Facility Management Practices

Empirical Cases in Developing Countries

Springer

Editors
Solomon Pelumi Akinbogun
Department of Built Environment
Faculty of Engineering
Built Environment and Information Technology
Biomimicry and Nature-Based ICT Solutions for Sustainable Construction Research Centre
Walter Sisulu University
Butterworth, South Africa

Clinton Ohis Aigbavboa
Department of Construction Management and Quantity Surveying
Faculty of Engineering and Built Environment
University of Johannesburg
Johannesburg, Gauteng
South Africa

Oluwaseun Tope Akinbogun
Department of Operation Management
Faculty of Engineering and Built Environment
University of Johannesburg
Johannesburg, South Africa

ISBN 978-3-031-65320-9 ISBN 978-3-031-65321-6 (eBook)
https://doi.org/10.1007/978-3-031-65321-6

This Springer imprint is published by the registered company Springer Nature Switzerland AG
The registered company address is: Gewerbestrasse 11, 6330 Cham, Switzerland

Disclaimer

The views expressed in the chapters of this book are those of the respective authors and do not necessarily reflect the views of the editor or the publisher. While every effort has been made to ensure the accuracy and completeness of the information contained in this book, the editor and publisher assume no responsibility for errors, omissions, or inaccuracies. The authors of each chapter are solely responsible for the content and any potential errors within their respective contributions.

About This Book

Human activities in every economy are driven by facilities which must be effectively managed for optimum performance. The management of facility in the public and private sector in the third world countries calls for thorough examination to explain the reasons for breakdown. From the economist point of view, some facilities are better provided by public sector to maintain inclusivity and protect the social welfare of the vulnerable people. In Nigeria, a good number of public facilities such as school, health facilities are under-performing due to poor and ineffective management. This paved the way for the private sector, however, a majority of the population are deprived due to affordability problem. This book explores the management of public facilities and the implications drawing context from educational, healthcare facilities, public spaces, and construction. These were essentially carried out to assess what could be done to ensure that essential services are not left in the hand of capitalist and possible synergy between the two sectors. This book is highly essential as the second- and third-order effect of the prevailing management of public facilities on service users are something to ponder. The book is useful to policy makers at all levels of the public and private sector.

Contents

List of Reviewers

We wish to acknowledge all our esteemed reviewers for their invaluable contributions to all the submissions. Specifically, the contributions of the following academics are highly appreciated.

Prof. Prof. Michael White, Real Estate Unit, School of Architecture Design and the Built Environment, Nothingham Trent University, UK

Prof. Oladokun Michael, Praetorian Construction Management, Edmonton Canada

Prof. Nicola Livingstone, Urban Studies, University of Glasgow, UK

Dr. Oyedokun Tunbosun, Urban Studies, University of Glasgow, UK

Prof. Gumbo Trynos, Department of Town and Regional Planning, University of Johannesburg, South Africa

Dr. Oke Ayodeji, Department of Quantity Surveying, Federal University of Technology, Akure, Nigeria.

Dr. Ikediashi Dubem, School of Computing and the Built Environment, Edinburgh Napier University, UK.

Prof. Emoh Fidelis, Real Estate, Department of Architecture and Planning, University of Botswana.

Prof. Bankole Awuzie, Faculty of Engineering and Built Environment, University of Johannesburg, South Africa.

Dr. N. D. Mbazor, Department of Estate Management, Federal University of Technology, Akure, Nigeria.

Dr. Oladirin Timothy, Dr. Olugbenga Oladinrin, School of Art, Design and Architecture (Faculty of Arts, Humanities and Business) University of Plymouth, UK

Dr. Makanju Sina, Department of Building, Federal University of Technology, Akure, Nigeria

About the Editors

Dr. Solomon Pelumi Akinbogun is an academic and the sub-dean at the School of Environmental Technology, Federal University of Technology, Akure, Nigeria. In addition, he is a research fellow at Walter Sisulu University in South Africa. He obtained his Ph.D. in June 2015 from the Institute for Housing, Urban and Real Estate Research at the School of Energy, Geoscience, Infrastructure, and Society (formerly known as the School of Built Environment), Heriot-Watt University, Edinburgh, UK. Prior to that, he earned his B.Tech. (Hons.) and M.Tech. in Estate Management from the Federal University of Technology, Akure, Nigeria, in 2001 and 2007. His research focuses on real estate economics, property and facility management, providing a narrative on the implications of current practices on sustainable development, property markets, and values. He has published several articles in top journals on these topics. He is also a reviewer for reputable refereed journals such as the *International Journal of Housing Market and Analysis* and the *Journal of Facilities Management*. Additionally, he is a member of the Society of Property Researchers in the UK and the Nigerian Institution for Estate Surveyors and Valuers.

Clinton Ohis Aigbavboa is a professor at the Department of Construction Management and Quantity Surveying, University of Johannesburg, South Africa. Before joining academia, he was involved as a quantity surveyor on several infrastructural projects, both in Nigeria and South Africa. He is the immediate past vice-dean of the Faculty of Engineering and Built Environment, University of Johannesburg, South Africa. He is a strong academic with extensive knowledge in practice, research, training and teaching. He is currently the chair of Sustainable Human Settlement and Construction Research Centre at the University of Johannesburg. He is also an author of five research books that were published with Springer Nature and CRC Press. He is currently the editor of the *Journal of Construction Project Management and Innovation* (accredited by the DoHET) and has received national and international recognition in his field of research. The South Africa National Research Foundation rates him as a Young researcher with the potential of establishing himself within a 5-year period.

Dr. Oluwaseun Tope Akinbogun is a lecturer at the Federal University of Technology, Akure, Nigeria. She earned her Ph.D. in Operations from the University of Johannesburg, South Africa. She also holds an M.Sc. in Strategic Project Management from Heriot-Watt University, Edinburgh, and a B.Sc. in Business Administration from Lagos State University. She has a robust portfolio of publications in reputable local and international journals, showcasing her expertise and contributions to the field.

Chapter 1
Facility Maintenance Culture in Nigerian Universities: A Case Study of a Public University

V. O. Ige and A. O. Asaju

Abstract This research examined the culture of facility maintenance at a Nigerian public University, focusing on evaluating maintenance strategies and the factors that influence the maintenance culture. Data was collected through close-end questionnaires administered to the University's works department. For data analysis, descriptive statistical methods include the relative importance index, weighted mean, and frequency distribution table. The findings indicate that key factors influencing the maintenance culture at the University include delayed responses to maintenance requests, insufficient funding for maintenance activities, and the absence of a maintenance policy. Reactive and corrective maintenance strategies and occasional renovations were identified as the predominant approaches employed at the University. Consequently, this study proposes that adequate funding should be allocated to the maintenance department to ensure prompt response to maintenance requests, thereby reducing the time required for repairs and preventing further deterioration. Moreover, the existing maintenance strategies at the University are deemed inadequate; thus, it is recommended that a proactive preventive maintenance culture is adopted with purpose-driven plans and policies, along with outsourcing facility maintenance services.

Keywords Facility · Maintenance · Maintenance strategies · Maintenance culture · University

---

V. O. Ige (✉)
Federal University of Technology, Akure, Ondo State, Nigeria
e-mail: voige@futa.edu.ng

Biomimicry and Nature-Based ICT Solutions for Sustainable Construction Research Centre, Department of Built Environment, Faculty of Engineering, Built Environment and Information Technology, Walter Sisulu University, East London, South Africa

A. O. Asaju
Department of Estate Management, School of Environmental Technology, Federal University of Technology, Akure, Nigeria

S. P. Akinbogun et al. (eds.), *Facility Management practices*,
https://doi.org/10.1007/978-3-031-65321-6_1

1.1 Introduction

Maintenance culture entails consistently adhering to practices that preserve the condition of various entities such as structures, machinery, facilities, amenities, and other items. It encompasses individuals' mindsets, actions, and the significance they attribute to maintenance in their daily lives (Suwaibatul-Islamiah et al. 2012). The primary objective of regular maintenance is to extend the functionality and usefulness of facilities, as they tend to deteriorate with usage (Onyemenam 2014). Maintenance culture can be defined as the capacity of individuals or groups to consistently safeguard the valued aspects of their lives (Onyemenam 2014). The preservation of institutional amenities holds crucial importance for the well-being of end users, such as researchers, teaching staff, management personnel, clerical workers, and administrative staff (Oyenuga et al. 2012). These facilities play a significant role in the institution's day-to-day operations, and their condition reflects the overall state of the community's well-being (Lee and Wordsworth 2015). Neglecting facility maintenance leads to increased operating costs and resource loss. Failure to address maintenance issues promptly often results in higher financial costs required to restore facilities to their normal state (Jackson 2013).

Consistent maintenance operations are essential for tertiary educational institutions to prolong the useful life of their assets (Bastidas 2017). Preservation activities should be an ongoing effort to maintain University facilities to a high standard, especially in the face of disasters (Bastidas 2017). Regular inspections of facilities are crucial in maintaining University assets, as they help identify faults before they develop into costly maintenance operations (Simpeh et al. 2023). African countries generally lack a maintenance culture (Osunrinde and Agbetuyi 2018). Nigerians, in particular, do not prioritise maintenance in their daily lives, whether in their residences, workplaces, educational institutions, or other organisations (Mbamali 2013). Many higher education institutions in developing nations neglect asset maintenance and lack structured guidelines to address depreciating assets (Oyenuga et al. 2012). For example, facilities at the University of Malawi are deteriorating due to a poor maintenance attitude, with inadequate funding being a major contributing factor as perceived by the institution (Oyenuga et al. 2012). Establishing proper maintenance procedures and guidelines for buildings poses significant risks to occupants and results in substantial financial losses (Ugwu et al. 2018). Numerous facilities in Nigerian public educational institutions are under-maintained, and even those that receive maintenance eventually deteriorate due to the lack of long-term maintenance plans (Ugwu et al. 2018).

Consequently, facilities in public educational institutions are rapidly depreciating, posing risks to the lives of those utilising them (Ohaedeghasi et al. 2021). Previous research has highlighted the ineffectiveness of facility maintenance in various Nigerian public tertiary institutions (Ohaedeghasi et al. 2021). Facility maintenance is crucial in extending the useful life of University facilities, and evidence suggests that the condition of facilities can impact students' performance (Osazuwa et al.

2021). In recent years, the maintenance culture of University facilities has gained attention from scholars, particularly in the developing world.

This study aims to analyse the culture of facility maintenance at the pseudonymous institution, University X, specifically focusing on the significance of maintenance practices and their impact on the institution. Using a pseudonym, "University X," is necessary to maintain the confidentiality and privacy of the institution being studied. Using a pseudonym ensures that the institution's identity remains undisclosed, protecting its reputation and preventing any potential biases or negative implications that might arise from the study's findings. The research aims to focus on the broader context of facility maintenance culture in Nigerian Universities rather than singling out and potentially stigmatising a specific institution. Therefore, using a pseudonym allows for an objective analysis of the maintenance culture without compromising the privacy of any particular University. Also, the selection of University X as the case study institution is motivated by several factors. Firstly, University X represents a typical Nigerian public tertiary institution facing maintenance culture challenges, as highlighted by previous research (Ohaedeghasi et al. 2021). Secondly, University X provides a suitable setting to investigate maintenance practices in an educational institution and their implications. By selecting University X as the case study institution, this research aims to shed light on the maintenance practices and factors that shape the maintenance culture at the campus. The findings of this study will contribute to the existing body of knowledge in facility management and provide valuable recommendations for decision-makers and practitioners seeking to improve maintenance techniques in educational institutions.

Studies conducted on the maintenance culture at Lagos State University have revealed issues related to the lack of planning and structured maintenance policies (Oyenuga et al. 2012), which align with findings from other studies (Ugwu et al. 2018). Additionally, research examining the maintenance culture of selected universities and polytechnic institutions found that inexperienced personnel often staff the maintenance departments (Olatunji et al. 2016). Studies on established university academic libraries have indicated a lack of training, inadequate funding, and insufficient retraining as obstacles to maintaining culture (Osunrinde and Agbetuyi 2018). Similarly, assessments of facility maintenance culture in selected northeast senior secondary schools have revealed poor maintenance practices (Garba 2020). The main motivation for conducting this study is to address the existing gaps in understanding regarding maintenance practices and the factors influencing the maintenance culture at University X. This will, in turn, provide insights into establishing a thriving facility maintenance culture and implementing effective maintenance procedures. The outcomes of this research will contribute to the existing body of knowledge in facility management and offer valuable recommendations for decision-makers and practitioners seeking to enhance maintenance techniques in educational institutions.

1.2 Maintenance Culture Concepts

The concept of instilling a maintenance culture among individuals, families, small organisations, businesses, and governments, wherein adherence to precise maintenance practices is emphasised, has been discussed by Ugwuzor et al. (2020). As defined by Suwaibatul Islamiah et al. (2012), maintenance culture encompasses the values, thought processes, actions, perceptions, and fundamental assumptions of individuals who consider maintenance essential and practice it daily. Establishing a maintenance culture ensures a deliberate effort to strive for continual improvement by focusing on structures that prevent obsolescence and enable adaptation to new technologies and changing norms (Ugwuzor et al. 2020). Consequently, a well-managed maintenance culture constantly reminds employees about the link between maintenance practices and the greater collective welfare.

1.2.1 Maintenance Strategies

The term "maintenance strategy" refers to the guiding principles that dictate scheduling tasks aimed at preserving facilities. This includes schedules for general maintenance, maintenance operations, and maintenance plans, as described by Akasah et al. (2016). Effective strategies, as noted by Waziri and Vanduhe (2013), involve clear objectives and proactive individuals capable of focus, adaptability, and adeptness in addressing unexpected challenges and ensuring security. Recent research has classified maintenance into planned and unplanned categories (Yates and Ge 2010). Planned maintenance entails organised and well-thought-out repairs to maintain buildings' currency, safety, and comfort over time. It can be categorised as preventive or corrective maintenance (Yates and Ge 2010). For instance, repairing a roof after a storm or replacing broken windows are unplanned or emergency maintenance examples. According to Zubairu (2015), neglecting maintenance in University buildings is unacceptable, as they require care to preserve their investment value, functionality, and aesthetic appeal. Various sectors and buildings employ a range of maintenance strategies, including breakdown maintenance, reactive maintenance, preventative maintenance, scheduled maintenance, predictive maintenance, condition-based maintenance, line maintenance, out-of-line maintenance, area maintenance, delayed maintenance, schedule-based maintenance, condition-based maintenance, performance-based maintenance, and breakdown-based maintenance, among others, as outlined by previous studies.

Shen et al. (2013) highlight that tertiary institutions face challenges in meeting the increased demand for maintenance planning. Gits (2014) suggests that proper maintenance planning and systematic inspection can prevent major maintenance emergencies. Simpeh et al. (2023) emphasise that maintenance cannot be effectively carried out without a plan. Budgeting for maintenance should align with the organisation's preventive maintenance strategy, as recommended by Gits (2014). According to the report, inadequate maintenance planning contributes to the deterioration of various infrastructures. Maintenance planning encompasses several factors that can reduce maintenance costs. Without planned maintenance, unexpected repairs can disrupt the life cycle of a University. Shohet and Straub (2013) assert that regular inspections are part of preventive maintenance. Their study proposes conducting condition surveys to assess the current state of facilities, identify areas requiring intervention, and prioritise maintenance activities. As stated in the study, preventive maintenance must incorporate inspection cycles until failure becomes unavoidable. Iyagba (2015) further emphasises that preventative maintenance is the most significant practice. The report recommends regular inspections to ensure optimal maintenance, requiring an understanding of the sources of degradation and the facilities themselves.

Additionally, a planned inspection schedule is advised to assess existing deterioration and provide recommendations for necessary maintenance planning work. For instance, the interval between two condition surveys should be specified. Gits (2014) explains that proper maintenance planning encompasses both preventive and remedial maintenance. Preventative maintenance ensures adequate spare parts and supplies, minimising repair costs while preserving the quality and capacity of machinery, structures, and other assets requiring maintenance. Senior management must recognise the critical nature of maintenance planning and the consequences of neglecting facility maintenance, as Simpeh et al. (2023) supported. Facility managers should be aware that all facilities necessitate maintenance plans. A single type of maintenance may not be suitable for every company.

According to AI-Najjar (2017), continuous and cost-effective maintenance enhances an organisation's profitability and competitiveness. Studies indicate that maintenance companies prefer reactive maintenance over proactive maintenance, resulting in under-maintained assets (El-Haram and Horner 2012). Under-maintained assets exhibit frequent and prolonged breakdowns, a high volume of unplanned tasks, and reduced productivity, as Anderson (2015) points out. Adopting a piecemeal approach to facility maintenance can lead to the complete collapse of the facility. Hence, planned preventative maintenance (PPM) was developed. PPM has been recognised as the most effective maintenance technique for minimising breakdown frequency (Wood 2013). However, Anderson (2015) argues that PPM can be ineffective as it may lead to premature and unnecessary replacements, fail to detect early signs of failure and incur significant costs.

Additionally, condition-based maintenance (CBM) is often called a maintenance strategy. CBM aims to generate maintenance recommendations based on condition monitoring data (Jardine et al. 2006). Consequently, the accuracy of CBM decisions depends on effective surveillance (Ayo-Imoru and Cilliers 2018). CBM focuses on controlling equipment failure modes, requiring consideration of all potential failure modes when implementing CBM. CBM operates under the assumption that most failures occur gradually and can be detected early in the deterioration process. The main challenge lies in determining when and how maintenance should be performed. Time-based maintenance is a long-established tradition, while TBM (time-based maintenance) plans maintenance based on a model that quantifies non-performance. It measures certain facilities' expected lifetime (T) using a non-performance period and prior information. TBM assumes characteristic failure behaviour for a facility (Guillén et al. 2016).

1.2.2 Influential Factors Affecting the Adoption of Maintenance Culture

Factors influencing maintenance culture encompass various elements that can act as barriers or drivers for successfully adopting maintenance practices. Several studies have highlighted these factors, shedding light on the challenges of effectively maintaining facilities. Some notable factors include the absence of a maintenance policy, inadequate financial resources, the presence of inexperienced and unqualified maintenance personnel, misuse of facilities by residents, and insufficient utilisation of technological advancements (Ugwu et al. 2018; Osunrinde and Agbetuyi 2018; Oyenuga et al. 2012; Olatunji et al. 2016; Ohaedeghasi et al. 2021).

Moreover, understanding the perspectives of facility users is crucial in evaluating maintenance practices. Waziri and Vanduhe (2013) conducted an evaluation study, considering the opinions of facility users through structured questionnaires. The findings emphasised the significance of maintenance culture. In Nigeria, poor maintenance culture prevails in private and governmental enterprises (Adejimi 2011; Usman et al. 2012). The absence of a maintenance policy has been associated with a deficiency in maintenance culture in Nigeria. However, conflicting views have been expressed by Olagunju (2012), who argued that inadequate funding for maintenance has also contributed to facility deterioration.

Similarly, Zubairu and Olagunju (2012) examined the maintenance of government office buildings using post-occupancy evaluation methods. Their study focused on deterioration in doors, ceiling boards, and windows. A lack of budgetary allocation for maintenance emerged as a contributing factor. Corruption has also been recognised as influencing the response to maintenance culture, as highlighted by Ugwuzor et al. (2020). When corruption levels are high, individuals may engage in maintenance activities only if they benefit from them. This vulnerability is particularly evident in government-owned property that lacks individual ownership. Consequently, neglect and mismanagement of equipment by government employees may occur, increasing the likelihood of illicit gains. In such corrupt situations, facilities and infrastructure may be left unattended for extended periods, treated as disposable, and disregarded by management and employees (Ugwuzor et al. 2020).

Ugwuzor et al. (2020) further identified factors such as a lack of awareness regarding the consequences of poor maintenance culture, an inadequate understanding of the importance of maintaining specific assets and processes, and noncompliance with maintenance instructions. Educational programs, orientation, and reorientation efforts are necessary for all individuals involved to address these issues. Encouraging repair, retrieval, and recycling practices can also contribute to developing a maintenance culture across all societal levels (Ugwuzor et al. 2020).

By comprehending and addressing these influential factors, decision-makers and practitioners can foster a more effective adoption of maintenance culture, leading to improved facility management practices.

1.3 Research Method

The data for this study was sourced from the work department of University X. This department was selected as the sample frame following the recommendation of Israel (2002), which suggests its suitability for small populations of 200 or fewer. The work department comprises four units: building, civil, transport and mechanical, and power supply. As of 2021, the department has 74 personnel (Staff Directory, University X). Of the 74 employees in the department, 60 individuals responded to the survey, accounting for a response rate of 61%. Various descriptive statistics were used to analyse the acquired data, including the Relative Importance Index (RII) and Weighted Mean Scores (WMS). The institution utilised the RII to evaluate the effectiveness of its maintenance methods, while the WMS was used to examine the University's facility maintenance culture.

The collected information was analysed through frequency tables and percentages to investigate the demographic characteristics of the work department's employees. The data collection instrument employed in this study was a questionnaire that consisted of items with five response options, ranging from "strongly agree" (5) to "strongly disagree" (1). Notably, "strongly agree" received the highest frequency of responses. The questionnaire consisted of three sections. Section A was designed to gather demographic information about the department's employees. Section B aimed to collect the data necessary for determining the department's maintenance strategies. Lastly, Section C comprised Likert-type questions concerning the factors influencing the University's facility maintenance culture.

1.3.1 Weighted Mean Score

To evaluate the influence of various factors on the safety of facility maintenance culture in the research region, a weighted mean score (WMS) was employed. The utilisation of WMS entails a statistical approach for assessing the significance and relative rankings of each aspect under consideration. The factors were arranged in descending order based on their weighted mean values, whereby the element with the highest weighted mean score was assigned the top rank, while the aspect with the lowest weighted mean score received the lowest rank.

For instance,

$$WM = (5\ 4\ 3\ 3\ 2\ 1)/5N \tag{1.1}$$

where:

n_5 "Strongly agreed" responses
n_4 number of "Agreed" responses
n_3 "Undecided" responses
n_2 "Disagreed" answers
n_1 Strongly Disagreed responses.

1.3.2 Relative Importance Index

The Relative Importance Index (RII) is a quantitative measure used to assess the relative importance of quality components. The Likert scale is employed to assign weights, denoted as W, to each factor based on the respondent's perception. The RII is calculated by considering the factor weighting given by the respondent. A higher RII indicates a greater weight given to the research. The total number of respondents, denoted as N, is used to determine the relevance of each factor. The RII values for different factors vary, enabling their ranking based on importance (Musa et al. 2014). In the present study, the RII was utilised to evaluate the diverse maintenance procedures employed within the research area.

$$RII = \sum W/A * N \tag{1.2}$$

Figure 1.1 depicts the organogram of the university's works department.

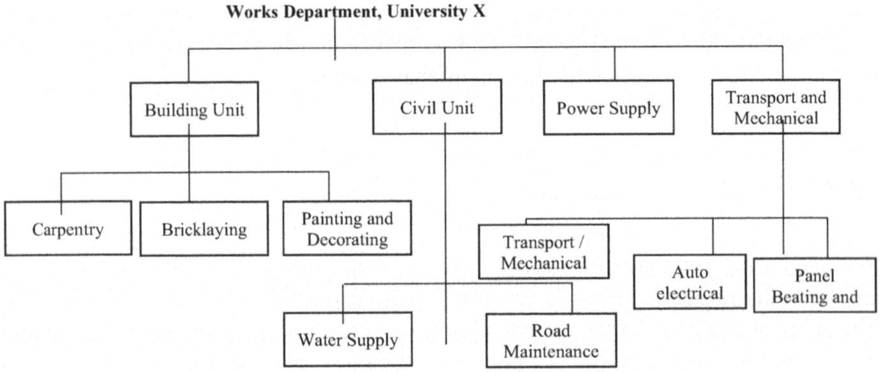

Fig. 1.1 Structure of Department of Works, University X. *Source* Field Survey, 2021

1.4 Analysis and Discussion

This section explores the data collected from the research area and presents the findings from the analysis. The primary objective of this analysis was to investigate the maintenance procedures employed at University X and understand the factors that shape the existing facility maintenance culture within the University. The analysis began by examining the maintenance practices at University X. Furthermore, the research discussed the relative importance of these maintenance strategies as perceived by the respondents. The Relative Importance Index (RII) was utilised to rank the maintenance strategies based on their significance. The analysis also considered the frequency of the maintenance strategies employed at University X. The results of this analysis provide valuable insights into the maintenance practices and culture at University X. By understanding the prevailing maintenance strategies and their frequency, decision-makers and practitioners can make informed decisions and develop targeted interventions to enhance the maintenance culture and optimise facility management within the University.

Table 1.1, displayed in the preceding section, illustrates the distribution of questionnaires among the maintenance personnel within University X's Department of Works. Out of the total of seventy-four (74) questionnaires disseminated to the department, sixty (60) were successfully collected, resulting in a response rate of 81%. Consequently, the data obtained from these respondents can be deemed reliable, enabling comprehensive analysis based on percentage figures.

The demographic characteristics of the respondents from University X, as classified by age, service length, department/unit, and qualification, are presented in Table 1.2. The age groups are delineated into five categories: 1–5 years, 6–10 years, 11–15 years, 16–20 years, and 21 years and above. The distribution of respondents across these categories indicates that 33.3% of the participants belong to the building unit, 20.0% are affiliated with the civil unit, 28.3% are associated with transportation and mechanical, and 18.3% are members of the power supply department. Regarding academic qualifications, most respondents possess a national diploma, a higher diploma, or a bachelor's degree, constituting 48.3% of the sample. A small percentage, 8.3%, holds a Master's degree, while no respondents reported having a PhD qualification. Additionally, 10.0% of the respondents indicated other qualifications beyond those mentioned above. This significant proportion of trained personnel reflects the proficiency and expertise of the respondents, reinforcing the

Table 1.1 Questionnaire administered and retrieved

Questionnaire	Frequency	Percentage (%)
Number distributed	74	100
Number retrieved	60	81
Number of un-retrieved	14	18

Source Field Survey, 2021

credibility and authenticity of the data obtained for this study. Importantly, University X demonstrates a higher representation of personnel with a bachelor's degree or a higher national diploma. This observation further bolsters the reliability and trustworthiness of the collected data and information, as the respondents possess the necessary educational background and training to provide valuable insights into the maintenance culture at the institution.

Table 1.3 provides valuable insights into the frequency of maintenance activities among the participants in the study. The data reveals that a significant

Table 1.2 Background information of respondents

Distribution	Frequency	Percentage (%)
Age of respondent		
18–24	0	0
25–30	0	0
31–36	5	8.3
37–42	11	18.3
43–48	10	16.6
49–54	15	25.0
55–60	13	21.6
61 years and above	6	10
Total	60	100
Length of service		
1–5	10	16.7
6–10	27	45
11–15	17	28.3
16–20	4	6.6
21 and above	2	3.3
Department/unit		
Building unit	20	33.3
Civil unit	12	20
Transport/Mechanical	17	28.3
Power supply	11	18.3
Academic qualification		
OND	10	16.7
HND/BSC	39	65.0
MSC	5	8.3
PHD	0	0
OTHERS	6	10
Total	60	100

Author's Survey, 2021

majority reported performing maintenance regularly, accounting for 86.7% of respondents. However, this finding implies that preventive maintenance practices are not commonly employed in the studied population. This finding aligns with previous research conducted in different contexts, indicating a consistent trend of infrequent implementation of preventive maintenance practices. For example, a study by Lavy and Bilbo (2009) focused on facility maintenance practices in educational institutions and found a similar pattern where preventive maintenance activities were reported as sporadic rather than regular.

Similarly, Fahmi et al. (2022) surveyed maintenance practices in manufacturing facilities and observed a lack of emphasis on preventive maintenance, indicating a prevailing trend of infrequent implementation. Furthermore, the current study's findings reinforce the notion that preventive maintenance practices are not prioritised in the maintenance culture of the studied population. This observation is consistent with a meta-analysis conducted by Eti et al. (2006), which synthesised findings from multiple studies and highlighted a widespread deficiency in adopting preventive maintenance strategies across various industries.

The findings presented in Table 1.4, which highlight the significant downtime experienced by facilities before repairs are carried out, have important implications for facility management. The data reveals that over 40% of the facilities examined in the study remained non-functional for more than two months before undergoing repairs. This finding aligns with previous research that has also identified maintenance response and resolution delays as a persistent issue in facility management. For instance, a study conducted by Ensafi and Thabet (2021), as referenced by Wuni et al. (2018), found similar trends in maintenance practices. The research revealed that many maintenance requests experienced substantial delays before being addressed, resulting in prolonged non-functional periods. This indicates that the problem of delayed repairs and extended downtime is not unique to the institution under investigation but is a prevalent concern in facility management across educational institutions.

Additionally, Abbas et al. (2019) researched maintenance practices in selected universities, and their findings also highlighted challenges related to the timely resolution of maintenance requests. The study discovered that lengthy delays in addressing maintenance issues were common, leading to prolonged impaired functionality for various facilities. These findings further reinforce the notion that the issue of delayed repairs and extended downtime are persistent problems within the institution and in the broader context of educational facilities. Considering the previous

Table 1.3 Frequency of responsiveness to maintenance activities

	No of occurrence	Percentage (%)	Cumulative percentage (%)
Periodically	52	86.7	86.7
Rarely	8	13.3	100.0
Total	60	100.0	

Source Field Survey, 2021

Table 1.4 Frequency of response to maintenance request

	Frequency	Percent (%)	Cumulative percent
One week	5	8.3	8.3
1–2 weeks	9	15.0	23.3
Within a month	10	16.7	40.0
1–2 months	11	18.3	58.3
More than two months	25	41.7	100.0

Source Field Survey, 2021

studies and the current findings in Table 1.4, it becomes evident that addressing the prolonged downtime experienced by facilities is an urgent issue. It underscores the need to develop and implement more efficient maintenance processes, proactive strategies for identifying and addressing maintenance issues, and adequate resource allocation to ensure timely and effective repairs. By addressing these challenges, educational institutions can create a more functional and conducive environment for their stakeholders, ultimately enhancing facility management practices' overall effectiveness and efficiency.

Table 1.5 presents the maintenance strategies employed in University X, their Relative Importance Index (RII) and corresponding rankings. The RII values range from 0.1502 to 0.0104, with higher values indicating a higher level of importance. Based on the table, the maintenance strategy with the highest ROI and a ranking 1 is "Corrective." This suggests that University X primarily relies on addressing maintenance issues as they occur, responding to problems reactively rather than proactively. The second-ranked strategy is "Based on Renovation," with an RII of 0.1155. This implies that renovation activities play a significant role in the maintenance practices of University X. The third-ranked strategy is "Preventive Planning," with an RII of 0.1042. This indicates that University X does employ some proactive maintenance measures to prevent issues before they occur, although they are not as highly prioritised as the corrective and renovation strategies. The remaining strategies, "Based on Condition," "Dependent on Time," "Dependent on Performance," "Based on Integration," and "Based on Breakdown," have lower RII values, indicating relatively lower importance or utilisation in University X's maintenance practices.

The findings of this study conducted at University X regarding maintenance strategies align with previous research in the field. Similar studies have emphasised promptly addressing maintenance issues to minimise disruptions and ensure operational efficiency (Oseghale 2014; Bagshaw 2017). This aligns with the first-ranked strategy, Corrective maintenance, which reflects the significance of reactive or corrective approaches. The second-ranked strategy, Based on Renovation, underscores the importance of regular upgrades and renovations in enhancing asset performance and extending their lifespan, as highlighted in prior research (Adenuga and Faremi 2012). Preventive Planned maintenance, ranked third, aligns with existing literature that advocates for planned maintenance activities to prevent failures and maintain optimal performance (Basri et al. 2017; Kundu et al. 2022). The relatively lower but still

Table 1.5 Maintenance strategies employed in University X

Maintenance strategies	RII	Ranking
Corrective	0.1502	1
Based on renovation	0.1155	2
Preventive planned	0.1042	3
Base on condition	0.0460	4
Dependent on time	0.0402	5
Dependent on performance	0.0188	6
Based on integration	0.0160	7
Based on breakdown	0.0104	8

Source Field Survey, 2021 (very low 1, low 2, average 3, high 4, very high 5)

significant importance of condition-based maintenance, ranked fourth, corresponds to the findings of previous studies that emphasise its value in proactively optimising maintenance activities (Meselhy et al. 2010; Mohideen and Ramachandran 2014). Time-based maintenance ranked fifth and has been recognised in prior research as a complementary approach for scheduling routine maintenance tasks at predetermined intervals (Chitra 2003; Javanmard and Koraeizadeh 2016). Furthermore, the lower perceived importance or less frequent usage of performance-dependent, integration-based, and breakdown-based strategies, ranked sixth to eighth, is consistent with the challenges or limitations associated with implementing these strategies effectively, as discussed in existing studies (Lee and Scott 2009; Velmurugan and Dhingra 2015; Deighton 2016).

The data obtained from the survey reveals that variables such as inadequate reaction time to maintenance culture, inadequate maintenance funding, and inadequate maintenance policy received the highest scores of 4.22, 4.10, and 3.96, respectively, indicating their greater significance. On the other hand, the factors of a lack of experts and qualified personnel in maintenance and ignorance of the consequences of an improper maintenance culture ranked fifth and sixth, with mean scores of 3.73 and 3.70, respectively. Therefore, the analysis of the Table 1.6 suggests that issues like a lack of responsiveness to maintenance requests, insufficient funding for maintenance, and the absence of a maintenance policy are substantial and relevant factors influencing the culture of facility maintenance at University X. Conversely, the factors of a lack of experts and qualified maintenance personnel and ignorance regarding the implications of an improper maintenance culture are not expected to have a significant impact.

This outcome is attributed to the prevailing belief among maintenance professionals that a lack of awareness regarding the implications of maintenance culture has a lesser influence on the maintenance culture of the University's facilities. These findings align with Ugwu et al. (2018) research, which identifies a lack of responsiveness to maintenance requests as a crucial factor influencing facility maintenance culture, but contradict the findings of Jandali et al. (2018), who identify a shortage of

Table 1.6 Factors affecting facilities maintenance culture in University X

Factor	SA(5)	A(4)	U(3)	D(2)	SD(1)	Mean score	Rank
Inadequate reaction time to maintenance culture	30(50.0)	17(15.0)	9(15.0)	4(6.6)	0	4.22	1*
Inadequate provision of funds for maintenance	28(46.7)	18(30.0)	10(16.7)	4(6.6)	0	4.10	2*
Lack of maintenance policy	24(40.0)	22(36.7)	10(16.7)	4(6.6)	0	3.96	3*
Corrupt administration	19(31.6)	20(33.3)	11(18.3)	10(16.6)	0	3.92	4
Lack of expert and qualified maintenance personnel	10(16.6)	14(23.3)	16(26.6)	20(33.3)	0	3.73	5
Ignorant as regards the implication of improper maintenance culture	9(33.5)	11(18.0)	15(33.9)	25(14.6)	0	3.70	6

Author's Survey, 2021 (the asterisk (*) represents the highest-ranked factor affecting the facilities maintenance culture in University X)
(Individuals who state that they "strongly agreed" are referred to as "S.A." Individuals who declare that they "agreed" are referred to as "A." Individuals who state "undecided" are referred to as "U.")

experienced and well-trained personnel as a significant element affecting the culture of facility maintenance in the institution.

1.5 Conclusion and Recommendation

A comprehensive examination was conducted on the facility maintenance culture at University X. The findings of this study highlight the absence of a maintenance policy and a deficient response system to maintenance requests within the University. It is advisable to allocate funds to the maintenance department to stock frequently utilised facilities and components to address these concerns. This measure will effectively enhance the institution's responsiveness to maintenance requests, minimise downtime, and mitigate further deterioration. Furthermore, the implementation of a purpose-driven preventive maintenance culture, accompanied by appropriate plans and policies, as well as the consideration of facility maintenance outsourcing, is strongly recommended. Lastly, further investigation into the facility's maintenance practices of high-rise buildings in renowned private Nigerian universities would be a valuable area for future research.

References

Abbas J, Aman J, Nurunnabi M, Bano S (2019) The impact of social media on learning behavior for sustainable education: evidence of students from selected universities in Pakistan. Sustainability 11(6):1–23. https://doi.org/10.3390/su11061683

Adejimi A (2011) Poor building maintenance: are architects free from blames? A paper presented at the ENHR international conference on housing: new challenges and innovations in tomorrow's cities Iceland, vol 170. pp 116–127

Adenuga AO, Faremi OJ (2012) Evaluation of maintenance management practice in banking industry in Lagos State, Nigeria. Int J Sustain Constr Eng Technol 3(1):45–53

Al-Najjar B (2017) The lack of maintenance and not maintenance which costs: a model to describe and quantify the impact of vibration-based maintenance on company's business. Int J Prod Econ 107:260–273

Akasah ZA, Shamsuddin SH, Rahman IA, Alias M (2016) School building maintenance strategy: a new management approach. In: Malaysian technical universities conference on engineering and technology (MUCEET). Universiti Malaysia Pahang, Pahang, pp 1–5

Anderson LM (2015) Spalling brick-material, design or construction problems. J Perform Constr Fac 163–171

Ayo-Imoru RM, Cilliers AC (2018) A survey of the state of condition-based maintenance (CBM) in the nuclear power industry. Ann Nucl Energy 112:177–188

Bagshaw KB (2017) A review and analysis of manufacturing firms' plant maintenance and replacement strategies in Nigeria. Afr J Bus Manage 11(2):17–26

Basri EI, Abdul Razak IH, Ab-Samat H, Kamaruddin S (2017) Preventive maintenance (PM) planning: a review. J Qual Maint Eng 23(2):114–143. https://doi.org/10.1108/JQME-04-2016-0014

Bastidas P (2017) Maintenance manual for school buildings in the Caribbean (online). Organization of American States, America. Available from http://www.oas.org/CDMP/documents/schools/maintman.htm. Accessed on 11 Dec 2021

Chitra T (2003) 'Life-based maintenance policy for minimum cost'. In: Annual reliability and maintainability symposium. pp 470–474. https://doi.org/10.1109/RAMS.2003.1182034

Deighton MG (2016) Maintenance management in facility integrity management. In: Deighton MG (ed) Facility integrity management. Gulf Professional Publishing, pp 87–139. https://doi.org/10.1016/B978-0-12-801764-7.00005-X

El-Haram MA, Horner MW (2012) Factors affecting housing maintenance cost. J Qual Maint Eng 8(2):115–123

Ensafi M, Thabet W (2021) Challenges and gaps in facility maintenance practices. 2:237–245. https://doi.org/10.29007/1h2j.

Eti MC, Ogaji SOT, Probert SD (2006) Development and implementation of preventive maintenance practices in Nigerian industries. Appl Energy 83(10):1163–1179. https://doi.org/10.1016/j.apenergy.2006.01.001

Fahmi RP, Annisa K, Asrie P, Dini DT, Rahmat N (2022) Implementation of maintenance management at manufacturing industry: a survey of maintenance characteristics. In: Proceedings of the international conference on industrial engineering and operations management Istanbul, Turkey, Mar 7–10, IEOM Society International

Garba MH (2020) Assessment of school facilities maintenance in Nigeria for the effectiveness of teaching process using North Eastern public senior secondary schools. 41(8):61–67

Gits CW (2014) Structuring maintenance control systems. Int J Oper Prod Manag 14(7):5–17

Guillén AJ, Crespo A, Gómez JF, Sanz MD (2016) A framework for effective management of condition-based maintenance programs in the context of industrial development of E-maintenance strategies. Comput Ind 82:170–185
Israel GD (2002) Sampling the evidence of extension program impact. Program Evaluation and Organizational Development, IFAS, University of Florida
Iyagba ROA (2015) The menace of sick building: a challenge to all for its prevention and treatment. Inaugural lecture series. University of Lagos Press, Lagos
Jackson E (2013) Choosing a methodology: philosophical underpinning. Pract Res High Educ J 7(1)
Jandali D, Sweis RJ, Alawneh AR (2018) Factors affecting maintenance of hospital buildings: a literature review. Int J Inf Bus Manage 10(3):69–83
Jardine AKS, Lin D, Banjevic D (2006) A review on machinery diagnostics and prognostics implementing condition-based maintenance. Mech Syst Signal Process 20:1483–1510
Javanmard H, Koraeizadeh A (2016) Optimising the preventive maintenance scheduling by genetic algorithm based on cost and reliability in National Iranian Drilling Company. J Ind Eng Int 12:509–516. https://doi.org/10.1007/s40092-016-0155-9
Kundu K, Cifone F, Costa F, Portioli-Staudacher A, Rossini M (2022) An evaluation of preventive maintenance framework in an Italian manufacturing company. J Qual Maint Eng 28(1):37–57. https://doi.org/10.1108/JQME-02-2020-0007
Lavy S, Bilbo DL (2009) Facilities maintenance management practices in large public schools, Texas. Facilities 27(1/2):5–20. https://doi.org/10.1108/02632770910923054
Lee H, Scott D (2009) Overview of maintenance strategy, acceptable maintenance standard and resources from a building maintenance operation perspective. J Build Apprais 4:269–278. https://doi.org/10.1057/jba.2008.46
Lee R, Wordsworth P (2015) Lee's building maintenance management. Blackwell Publishing, London
Mbamali P (2013) The impact of accumulation deferred maintenance on selected buildings of Two Federal Universities in the Northwest zone of Nigeria. J Environ Sci 3(17):2–4
Meselhy KT, Elmaraghy WH, Elmaraghy HA (2010) A periodicity metric for assessing maintenance strategies. CIRP J Manuf Sci Technol 3(2):135–141. http://dx.doi.org/10.1016/j.cirpj.2010.06.004
Mohideen PBA, Ramachandran M (2014) A strategic approach to breakdown maintenance on construction plant—UAE perspective. Benchmarking Int J 21(2):226–252. https://doi.org/10.1108/BIJ-05-2012-0030
Musa, Mohammad MF, Mahbub R, Yusof MR (2014) Soc Behav Sci 153:79
Ohaedeghasi CI, Ezeokoli FO, Agu NN, Okolie KC, Ugochukwu SC (2021) Prospects for improving building maintenance management in Nigerian Public Universities; case study of Nnamdi Azikwe University. J Eng Res Rep 21(4):53–59
Olagunju RE (2012) Predictive modelling for sustainable residential building maintenance in developing countries: a Nigerian case. Interdisc J Contemp Res Bus 4(6):1237–1283
Olatunji SO, Aghimien DO, Oke AE, Akinkunmi T (2016) Assessment of maintenance culture in Tertiary Institutions in Nigeria. Civ Environ Res (Online) 8(6):101–105
Onyemenam CT (2014) Maintenance culture in Nigeria: an intuition into the Nigerian postal services. ArticlesNG
Osazuwa ON, Iroham CO, Oluwunmi AO (2021) Factors affecting the effectiveness of maintenance in postgraduate hostels in highly ranked Nigeria universities. In: IOP conference series: earth and environmental science, vol 655
Oseghale GE (2014) Impact of maintenance strategies on the performance of industrial facilities in selected industrial estates in Lagos State, Nigeria. Am J Eng Res (AJER) 3(8):171–179
Osunrinde AA, Agbetuyi PA (2018) Maintenance culture in academic libraries of Universities in Southwest Nigeria. Int J Libr Inf Sci Stud 4(4):10–22
Oyenuga SO, Akinsola OE, Hussain PO, Fatokun AO (2012) Maintenance culture on University facilities in Lagos State University, Ojo, Nigeria as a case study. Mediterr J Soc Sci 3(11):97–99

Shen QP, Lo KK, Wang Q (2013) Priority setting maintenance management of public buildings—a modified multi-attribute approach using analytic hierarchy process. J Constr Manage Econ 16(6):693–702

Shohet IM, Straub A (2013) Performance-based-maintenance: a comparative study between the Netherlands and Israel. Int J Strateg Prop Manag 17(2):199–209

Simpeh F, Mavangwa GM, Wentzel L (2023) Strategies to overcome barriers to public facilities management. Prop Manag 41(3):365–380. https://doi.org/10.1108/PM-03-2022-0016

Suwaibatul Islamiah AS, Abdul Hakim M, Syazwina FAS, Eizzatul AS (2012) An overview development of maintenance culture. In: Proceeding conference 3rd international conference on business and economic research, pp 2206–2217

Ugwu OO, Okafor CC, Nwoji CU (2018) Assessment of building maintenance in Nigerian University system: a case study of the University of Nigeria, Nsukka. Nig J Tech 37(1):44

Ugwuzor M, Okpu TO, Diett-spiff (2020) Maintenance culture management for global impact. J Resour Dev Manage 64.https://doi.org/10.7176/JRDM/64-04

Usman ND, Gambo MJ, Chen JA (2012) Maintenance culture and its impact on the construction of residential buildings in Nigeria. J Environ Sci Resour Manage 12(6):501–507

Velmurugan RS, Dhingra T (2015) Maintenance strategy selection and its impact in maintenance function: a conceptual framework. Int J Oper Prod Manag 35(12):1622–1661. https://doi.org/10.1108/IJOPM-01-2014-0028

Waziri BS, Vanduhe BA (2013) Evaluation of factors affecting residential building maintenance in Nigeria: users' perspective. Civ Environ Res 3(8):19–24

Wood B (2013) Approaching the care free building. Facilities 21(3/4):74–79

Wuni IY, Agyeman-Yeboah S, Boafo HK (2018) Poor facility management in the public schools of Ghana. Recent empirical discoveries. J Sustain Dev Stud 11(1):1–30

Yates JM, Ge Z (2010) Network management: fault management, performance management, and planned maintenance. In: Guide to reliable internet services and applications. Springer, London, pp 397–445

Zubairu SN (2015) The most frequently recurring maintenance problems in government office buildings in Nigeria. NIAJ 11(136):8–12

Zubairu SN, Olagunju (2012) Post occupancy evaluation of some selected secondary schools in Minna, Nigeria. J Econ Sustain Dev (Online) 3(7):71–75

Chapter 2
Identifying the Skills and Competencies Required for Facility Management Roles: A Text Mining Study

Olalekan Shamsideen Oshodi, Rotimi Boluwatife Abidoye, Majeed Oladokun, and Abiodun Olatunji Abisuga

Abstract Due to the impact of facilities on the performance of the workforce, there has been a growth in the demand for facilities managers across the globe. To address this shortfall in skills, universities are creating various programmes to satisfy the needs of the industry. However, the skills and competencies required to succeed as facilities managers in the various sectors of the economy remains unclear. The current study seeks to address the gap in knowledge by uncovering the core and soft skills required for facility management job roles. Job advertisements are scrapped from webpages of a job board. The job adverts contain information about qualifications and essential skills required for facility management job roles. Text mining was used to analyse the collected data. Based on the frequency of mention, it was found that risk management, people management, operations and maintenance and procurement are the topfour core skills. Also, team building, problem-solving, communication and time management are the top soft skills required for facility management practice. The findings emerging from this study provides information that can be used for the development of the curriculum for facilities management programmes.

Keywords Data analytics · Education · Facility management · Graduate employability · Text mining

O. S. Oshodi (✉)
School of Engineering and the Built Environment, Anglia Ruskin University, Chelmsford, UK
e-mail: ooshodi1@gmail.com

R. B. Abidoye
School of Built Environment, UNSW, Sydney, Australia

M. Oladokun
School of Architecture, Building and Civil Engineering, Loughborough University, Loughborough, UK

A. O. Abisuga
Lincoln School of Architecture and the Built Environment, University of Lincoln, Lincoln, UK

S. P. Akinbogun et al. (eds.), *Facility Management practices*,
https://doi.org/10.1007/978-3-031-65321-6_2

2.1 Introduction

The construction industry creates the infrastructure needed to support economic growth. Over 40 years of research has shown that investment in infrastructure stimulates economic growth (Dang and Low 2011). Also, evidence from literature shows that the construction industry is linked to several sectors of the economy (Pietroforte and Gregori 2003). For instance, the construction industry utilises products of the manufacturing sector (such as tiles, cement and steel reinforcement, among others) to build new homes. Therefore, if there is a demand for new homes, it would lead to a growth in the production output of the manufacturing industry. In the developed world, the infrastructure and housing stock is ageing (Briggs et al. 2022), and this situation has led to the emergence of facilities management as an academic discipline.

As a new discipline, there are a few areas that are of interest to researchers. Some of the first questions asked include the need to distinguish facilities management from other built environment disciplines, such as construction management. For example, Azasu et al. (2018) argued that a lack of clear definition of skills and competencies required for facilities management jobs would make it difficult to create a curriculum for the training of students. Also, a limited understanding of the required competencies leads to a situation where the graduates of such programmes are not job-ready after their training (Aboagye and Puoza 2021). Therefore, a clear understanding of the skills and competencies required for facility management roles is vital for the development of effective education programmes.

Several studies have been conducted to define the skills and competencies required for facility management job roles. For example, Azasu et al. (2018) conducted a survey in South Africa to highlight the core and optional skills required for Facility management job roles. Priyangika et al. (2020) identify the skills needed by facilities managers in the hospitality sector. As stated earlier, facilities management is an academic discipline that is still evolving. Hence, there is a need to unearth the competencies required for facilities management job roles in the industry using the United Kingdom as a representative case. The current study seeks to address two questions: (i) What are the technical competencies required for facility management job roles? (ii) What are the soft skills required for facility management job roles? This information would help stakeholders gain an in-depth understanding of the core competencies and soft skills required for facilities management job roles. This is vital for the development of programmes targeted at improving the employability of graduates of facility management programmes.

2.2 Skills and Competencies Required for Success in Facility Management Job Roles

Several authors, stakeholders and professional bodies have tried to identify the core competencies required for the practice of facility management as a profession. A survey of facility management professionals by Azasu et al. (2018) showed that "customer service", "project management", "working with suppliers and specialists", "maintenance management", "corporate real estate management" and "procurement" are the core competencies of facility management. In contrast, Priyangika et al. (2020) found that "building services management", "project management", "space planning and management", "facility management strategy development" and "quality management" are the most significant competencies for facility management practice in the hospitality sector. Taken together, it is evident that the competencies required tend to vary based on the sector where the facility manager is practising.

Various professional bodies have also created competency framework for facility management practice. For instance, according to the International Facilities Management Association (2024), the skills needed by facilities managers are "finance and business", "operations and maintenance", "leadership and strategy", "project management", "communication", "performance and quality", "facility information management and technology management", "occupancy and human factors", "real estate", "risk management" and "sustainability". Also, the Royal Institute of Chartered Surveyors (2017) stated that the technical competencies required for facility management practice are: "analysis of client requirements", "building information modelling", "contract practice", "procurement and tendering", "project financial control and reporting", "property management" and "supplier management". On the overall, it seems that every stakeholder has different perspectives on the core skills needed to succeed as a facility manager. The current study seeks to unearth the key competencies through an analysis of jobs advertised by various employers.

2.3 Research Method

Several methods have been used to investigate problems in the field of facilities management and built environment. The methods adopted in previous research include: questionnaire survey (Azasu et al. 2018), interviews (Hou and Wu 2020), mixed method (Priyangika et al. 2020) and field observation (Hou and Wu 2020), among others. The present study utilises text mining to analyse data that was collected from online job board through the use of web scrapping. Chapagain (2019) stated that "webscraping is a process of data extraction from the web...". Webscraping is an automated process that eases the process of collecting large volumes of data from various webpages. Scholars, such as Wing et al. (1998), have stated that the suitability of a research method is dependent on the problem under investigation. A major advantage of text mining is that it allows researchers to analyse large volumes of unstructured data (such as social media posts, research reports and emails, among others) without the need to manually review the original text (Kim and Ju 2019; Moro et al. 2017). Also, text mining can be used to extract valuable insights from text which can be used for the development of forecasting models. Oliveira et al. (2017) used a text mining method, i.e., sentiment analysis, to analyse textual data from Twitter. The findings of the study showed that text is useful for predicting stock market behaviour. In this study, online job adverts were automatically scrapped from a job board. This data was analysed to understand the key competence for facility management job roles as required by employers.

To address the research questions, the study was carried out in four steps as described. Step 1—collect facility management-related job postings on an online job board. The search keyword used for the search is "facility manager". Most of the online job boards do not grant access to their data. On the day of data collection, it was only this job board that allowed the use of web scrapping on its webpage. The data was scrapped from the job board in October 2023. R programming language (R Core Team 2021) and the polite package were used for the data collection process, i.e., web scraping. In total, 9152 job adverts were scrapped from the job website. Step 2—data pre-processing and cleaning to remove inconsistent data, duplicates and anomalies (e.g., stop words removal). Step 3—text extraction: identify the presence of specific keywords (keywords are related to facility management competence) within the corpus of text. Step 4—create visuals. The collected data contained information about qualifications, essential skills and desired skills required for facility management job roles. These visualizations are useful for summarising and presenting text in an easy-to-understand format.

2.4 Results and Analysis

2.4.1 Unigram (Most Frequent Words in Facility Management Job Adverts)

Schonlau et al. (2017) stated that "a variable containing frequency counts of a single word in each text is called a unigram". Unigrams are useful for exploring the context of a text corpus. The most frequent words are summarised and presented in Fig. 2.1. As expected, management, facilities and experience are the most frequent words. It is also evident that facilities managers are involved in the delivery of excellent service which is linked to health and safety. Research has shown that the state of a facility affects the health of its users (Leung et al. 2017). It is expected that facility managers would use their knowledge to improve the quality of the indoor environment which affects the health and wellbeing of the users of these facilities.

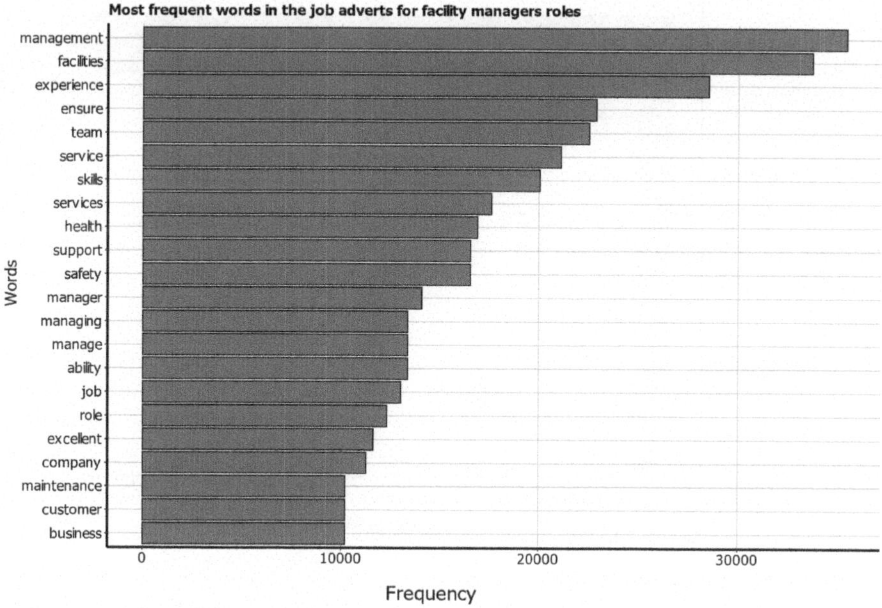

Fig. 2.1 Most frequent words used in the job adverts

2.4.2 Sectors Providing Facility Management Jobs

The data presented in Fig. 2.2 shows the sectors that employ those with the knowledge of facilities management. Services and recruitment agencies are the top employers of facility managers in the UK. This finding is not surprising because the economy of the UK is driven by services (Huggins et al. 2024). Hence, there are more jobs for facilities managers in this sector. Overall, it is evident that almost all the sectors of the economy need facility managers from health to social housing.

2.4.3 Qualification Required for Facility Management Roles

An analysis of the content of the job descriptions was done to understand the qualifications required for the practice of facility management (see Fig. 2.3). The findings indicate that most of the employers stated that they want applicants who possess a master's degree. It is worth noting that there was no mention of PhD in the qualifications required by employers of facility managers. The findings show that employers have a preference for holders of master's and bachelor's degrees for facility management job roles.

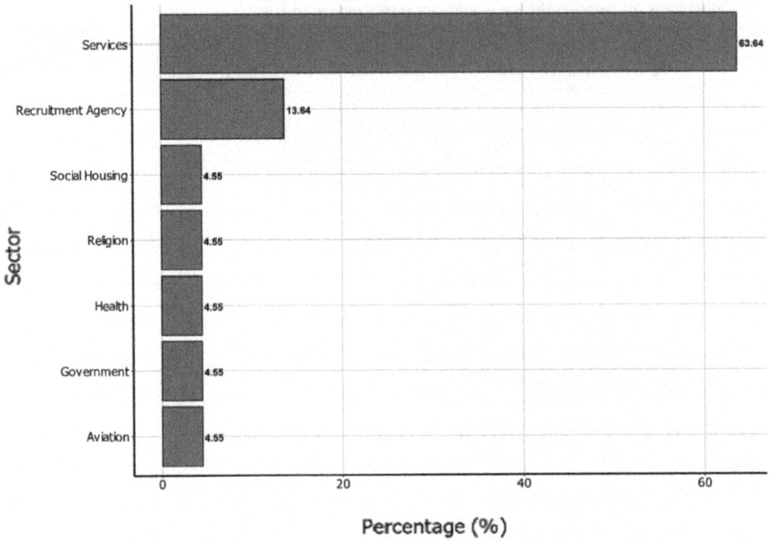

Fig. 2.2 Sectors that advertised facilities management job roles

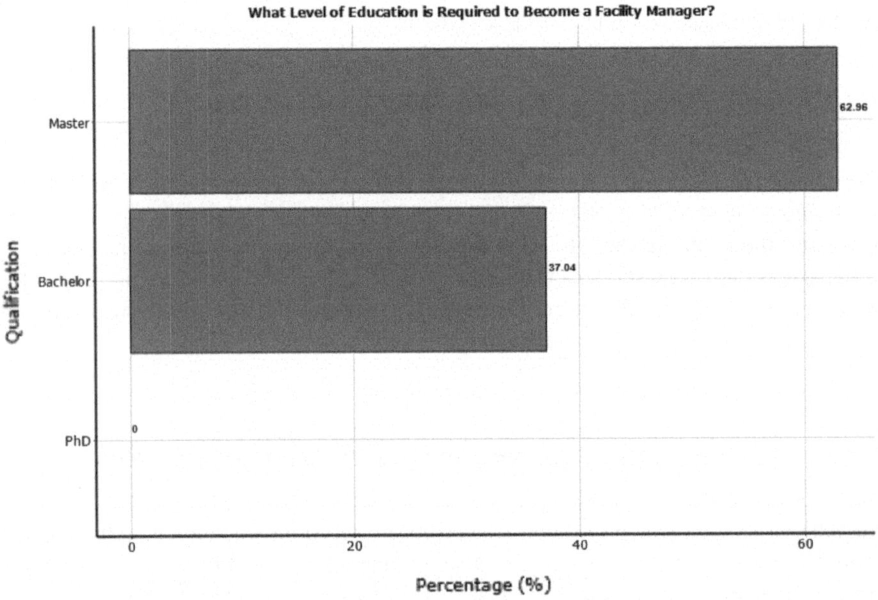

Fig. 2.3 Qualification required for facilities management roles

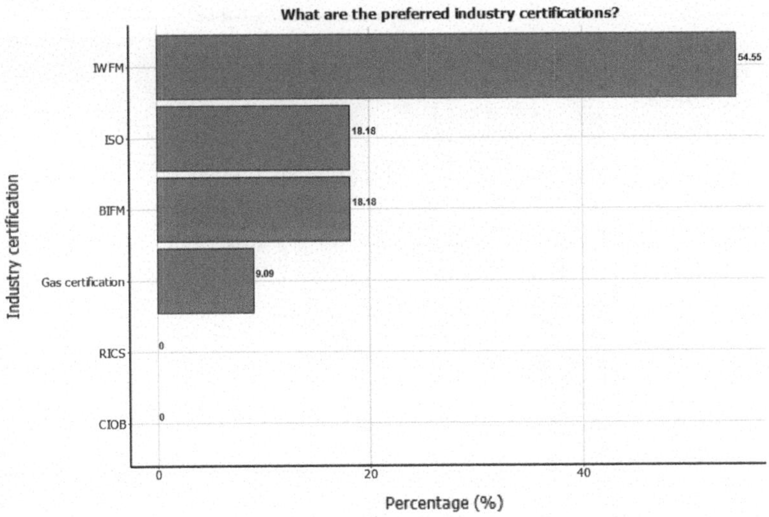

Fig. 2.4 Industry certifications required for facilities management job roles

2.4.4 *Industry Certifications*

Figure 2.4 shows the industry certifications needed by employers of facility managers as stated in the job advertised on online job boards. Figure 2.4 shows that the most frequently mentioned certification is Institute of Workplace and Facilities Management (IWFM) while gas certification (such as Gas Safe registered) is the least mentioned. Similarly, some employers stated that applicants should possess International Organization for Standardization [such as ISO 14001 and 45001] and British Institute of Facilities Management (BIFM) certifications. What is striking about the numbers presented in Fig. 2.4 is that there is no mention of the Royal Institution of Chartered Surveyors (RICS) and Chartered Institute of Building (CIOB) certifications. A likely reason for this observation is that RICS is an industry certification for building surveying and quantity surveying while CIOB is another established certification for construction managers and builders. Hence, as an emerging discipline, employers of facility managers require specific industry certifications.

2.4.5 *Technical Skill Set*

The results obtained from the analysis of the technical skill set that facility managers need to possess are shown in Fig. 2.5. The data presented in Fig. 2.5 shows that the top-four technical skills (based on the frequency of mention in job adverts) are risk management, people management, operations and maintenance, and procurement. While the bottom-four technical skills are supply chain management, property of

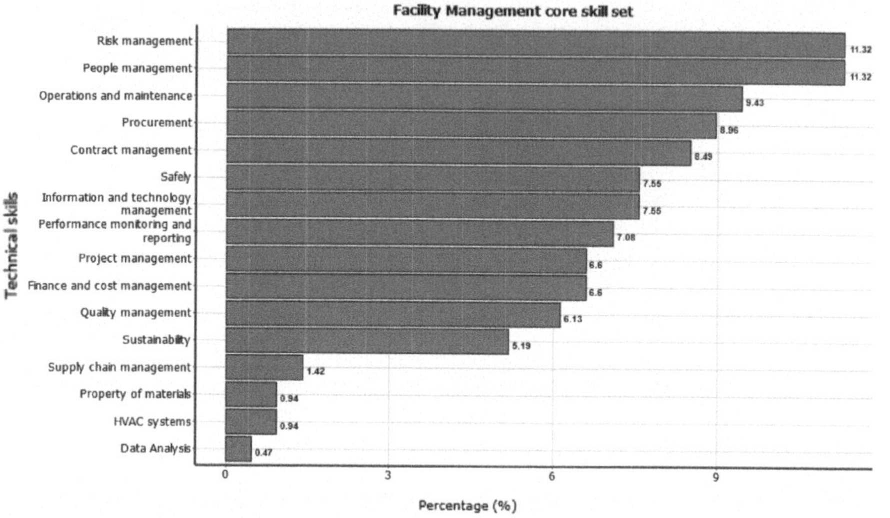

Fig. 2.5 Technical skills required for facility management job roles

materials, heating ventilation and air-conditioning (HVAC) systems, and data analysis. The most interesting aspect of this figure is that it shows that a wide variety of technical skills (including project management, safety and sustainability, among others) are needed to succeed as a facility manager. Also, due to the advances in communication and data technologies, it is not surprising that facility managers need to possess skills relating to information and technology management.

2.4.6 Soft Skills

Figure 2.6 shows the summary statistics for the soft skills needed for facility management job roles. The top-4 soft skills, based on the frequency of mention in job adverts, are team building, problem solving, communication and time management. In contrast, multi-tasking, decision making, presentation (public speaking) and active listening are the bottom-four soft skills stated in job adverts. From the data in Fig. 2.6, it is apparent that a multiplicity of soft skills are needed for facility management job roles. This is an expected result because facility managers are working with multiple stakeholders (such as vendors, clients, landlords and statutory agencies, among others) to ensure that facilities are performing optimally throughout of its usable lifespan.

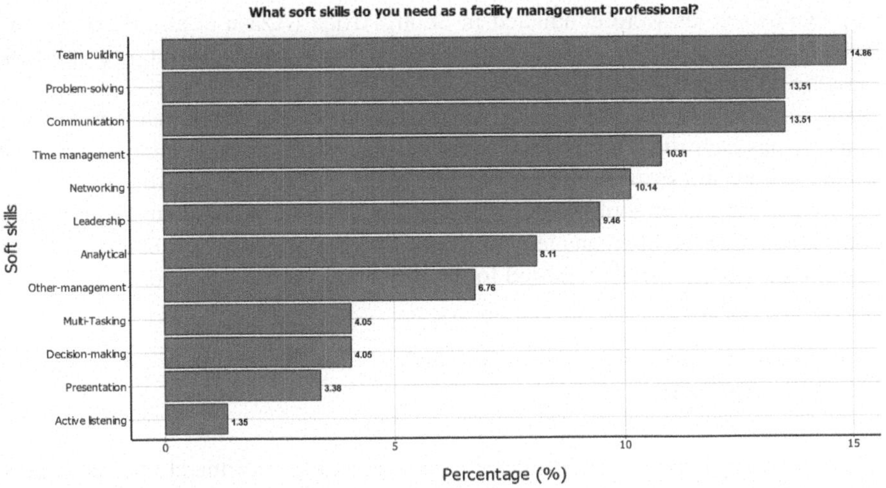

Fig. 2.6 Soft skills required for facility management job roles

2.5 Discussion

Prior studies have shown that facility management is an emerging academic discipline that plays an important role in keeping the economy productive as infrastructure ages. The present study was designed to unearth the skills and knowledge required for facility management job roles. In this study, it was found that risk management, people management, operations and maintenance and procurement are the top-four technical skills required for facility management job roles. Also, team building, problem solving, communication and time management are the top-four soft skills. The most unexpected finding is that CIOB and RICS were not mentioned in the job adverts for facility management job roles. This finding was unexpected and suggests that employers of facility managers are looking at facility management as a discipline that goes beyond just the management or maintenance of buildings.

The findings emerging from this study are consistent with those reported in similar previous research. For instance, Azasu et al. (2018) showed that "project management", "maintenance management", "procurement" and "customer service" are key technical skills required for the practice of project management. Also, the study aligns with the findings in Priyangika et al. (2020) who found that project management and quality management are key competencies required for the practice of facility management in the hospitality sector. In contrast to earlier findings (Azasu et al. 2018; Priyangika et al. 2020), no evidence of the mention of "corporate real estate management", "space planning and management" and "facility management strategy development" within the job adverts. There are several possible explanations for these observed inconsistencies. First, the previous studies were based on a questionnaire survey (Azasu et al. 2018) and mixed method (Priyangika et al. 2020) while the study reported in this paper is based on an analysis of job adverts. Second, the previous studies were conducted in South Africa (Azasu et al. 2018) and Sri Lanka (Azasu et al. 2018) while this study is based on jobs advertised by employers operating within the United Kingdom.

The findings of the current study reported in this paper show that some of the competencies identified in this study were not mentioned in previous research. For instance, safety and sustainability are unique findings that emerged from the current study. These variations could be due to the emergence of new competencies required for the practice of facility management. These findings provide some evidence about the technical and soft skills required for facility management practice.

2.6 Conclusions

The main goal of the current study is to understand the technical and soft skills required for the practice of facilities management. The study showed that risk management, people management, operations and maintenance and procurement are the top-four technical skills. The second major finding was that team building, problem-solving, communication and time management are the top-four soft skills. In terms of industry certifications, IWFM, ISO, BIFM and gas safe are the top-four industry certifications required for facility management job roles. Taken together, these results show that a wide variety of technical and soft skills are required to succeed in facility management job roles.

The findings of this investigation complement those of earlier studies by highlighting new skills and competencies that were not previously stated. Overall, this study suggests that facility management is still emerging, and the key skills needed for its improvement are changing to align with the demands of the job role. In the future, the use of technologies, such as digital twins or internet of things, could emerge as key skills needed for facility management. The findings reported here shed new light on the core and soft skills required for facility management practice.

Being limited to job adverts published by employers operating in the United Kingdom, the findings are less generalisable to other countries or contexts. Also, it is possible that some of the skills stated in job adverts are not really required for facility management job roles. These are limitations of the results emerging from the study. In spite of its limitations, this study certainly adds to our understanding of the core and soft skills needed for securing facility management job roles. Further research using alternative methods, such as ethnography, should be carried out to establish the skill set needed by facility managers through observing these professionals doing their day-to-day jobs. An understanding of the required skill sets would help universities to develop programmes that improve the employability of their graduates.

References

Aboagye B, Puoza JC (2021) Employability of mechanical engineering graduates from Sunyani Technical University of Ghana. J Teach Learn Graduate Employability 12(2):185–205

Azasu S, Adewunmi Y, Babatunde O (2018) South African stakeholder views of the competency requirements of facilities management graduates. Int J Strateg Prop Manag 22(6):471–478

Briggs KM, Dijkstra TA, Glendinning S (2022) Assessing the deterioration of ageing infrastructure earthworks. GeoStrata Mag Arch 26(4):26–33

Chapagain A (2019) Hands-on web scraping with Python. Packt Publishing Limited

Dang THG, Low SP (2011) Role of construction in economic development: review of key concepts in the past 40 years. Habitat Int 35(1):118–125

Hou H, Wu H (2020) A case study of facilities management for heritage building revitalisation. Facilities 38(3/4):201–217

Huggins R, Munday M, Thompson P, Xu C (2024) Entrepreneurial ecosystems, agency and regional development: emergence and new path creation in the Cardiff city region. Local Econ 02690942241237779

International Facilities Management Association (2024) What is facility management? Retrieved 2 Feb 2024 from https://www.ifma.org/about/what-is-fm/#c

Kim L, Ju J (2019) Can media forecast technological progress?: A text-mining approach to the on-line newspaper and blog's representation of prospective industrial technologies. Inf Process Manage 56(4):1506–1525

Leung M-Y, Famakin I, Kwok T (2017) Relationships between indoor facilities management components and elderly people's quality of life: a study of private domestic buildings. Habitat Int 66:13–23

Moro S, Rita P, Cortez P (2017) A text mining approach to analyzing annals literature. 208–210

Oliveira N, Cortez P, Areal N (2017) The impact of microblogging data for stock market prediction: using Twitter to predict returns, volatility, trading volume and survey sentiment indices. Expert Syst Appl 73:125–144

Pietroforte R, Gregori T (2003) An input-output analysis of the construction sector in highly developed economies. Constr Manag Econ 21(3):319–327

Priyangika K, Perera B, Wickremanayake Karunaratne TL (2020) Facilities management roles in the hotel industry: the skills and competencies required. J Qual Assur Hosp Tour 21(4):454–473

R Core Team (2021) R: a language and environment for statistical computing. R Foundation for Statistical Computing. https://www.R-project.org/

Royal Institute of Chartered Surveyors (2017) Associate Assessment: Facilities Management. https://www.rics.org/content/dam/ricsglobal/documents/join-rics/facilities_management_pathway_guide_associate_rics.pdf

Schonlau M, Guenther N, Sucholutsky I (2017) Text mining with n-gram variables. Stata J 17(4):866–881

Wing CK, Raftery J, Walker A (1998) The baby and the bathwater: research methods in construction management. Constr Manag Econ 16(1):99–104

Chapter 3
The Fourth Industrial Revolution and Digitalisation of Facilities Management Firms in Nigeria

Henry Okpo Asuquo and Michael Gbolagade Oladokun

Abstract This study explored the critical context-based issues fostering low adoption of digitalisation transformation in Nigerian facilities management (FM) firms. The study adopted a qualitative research approach through interviews, carried out among eight case FM firms in Nigeria purposively selected. The study utilised thematic analysis in analysing data generated. The study revealed the most significant factors inhibiting the adoption of digital transformation of FM firms, among which are lack of funds, shortage of trained professionals, lack of technical know-how, and low level of awareness. Also, the study revealed the strategies aiding the digital transformation of FM firms as government support, digital collaboration, training, increase in research and development, and standardisation of the industry using Building Information Modelling. The study concluded that the strategies aiding the digital transformation of FM firms are effective, as revealed by the case study firms, with some noticeable differences. The study recommended collaboration between government and stakeholders in the FM industry to work out an appropriate roadmap that would position the FM industry on the right track towards the adoption of the fourth industrial revolution and digitalisation in Nigeria.

Keywords Digitalisation · Facilities management · Fourth industrial revolution · Nigeria

H. O. Asuquo
Department of Building, Faculty of Environmental Studies, University of Uyo, Uyo, Nigeria

M. G. Oladokun (✉)
Department of Building, Faculty of Environmental Design and Management, Federal University Oye-Ekiti, Oye, Ekiti State, Nigeria
e-mail: michael.oladokun@fuoye.edu.ng

S. P. Akinbogun et al. (eds.), *Facility Management practices*,
https://doi.org/10.1007/978-3-031-65321-6_3

3.1 Introduction

According to its definition, facility management (FM) is "a profession that integrates people, places, processes, and technology to assure the operation of the physical environment" (Wang et al. 2022). Araszkiewicz (2017) posited that management activities in the core areas of communication and information management are crucial at every step of the life cycle of a building project. For a building project, the planning and design phase normally lasts two to three years, whereas the operation and maintenance phase frequently lasts decades. The majority of life cycle costs 2/3, or $15.8 billion were expended in facilities management or operations and maintenance (Wang et al. 2022). The phase of a facility or construction project that consumes the most time, money, and energy, however, is the one that involves operation and maintenance. Therefore, the processes of these operations need to be revolutionised. The fourth industrial revolution focuses on digitalisation to cut costs and reduce energy use. A way for an organisation to deliver and maintain its buildings and auxiliary facilities to meet strategic objectives and requirements under changing circumstances is known as facilities management (Fadahunsi et al. 2019). The fourth industrial revolution has rapidly changed the practice and application of facilities management over the past ten years. This is partially due to the growing emphasis on more user-friendly, customised premises as FM businesses start to recognise the benefits of the idea. The planning and execution of a building project may be improved using cutting-edge digital technologies and tools and doing so makes it possible to compile the data needed to operate the structure effectively (Araszkiewicz 2017).

The coin facility 4.0 used in this investigation originates from the fourth industrial revolution. In many organisations, Facility 4.0 brought digitalisation as a top priority. Facility 4.0 refers to the rising digitisation, connectivity, and commercial practices of FM goods, services, value chains, and processes. IoT (internet of things), augmented reality, big data, artificial intelligence, and other cutting-edge technologies are all combined to power facility 4.0. Facility 4.0 is motivated by the need for the physical and virtual worlds to merge; the following stage combines traditional and cutting-edge technology to provide FM service. It is frequently incorporated into organisational strategies, policies, objectives, goals, structures, and responsibilities, as well as projects and investment plans, with implicit expectations of industry reforms and greater overall efficiency (European Commission 2018; World Economic Forum 2019; The World Bank Group 2019; OECD 2019). Facility 4.0 is linked to other significant changing movements that have a profound effect on people's day-to-day lives in society and the advancement of society, such as digitalisation and sustainability, among others (European Commission 2018). The essential characteristics of digital technologies are what spark interest in facility 4.0.

The industrial revolution started at the end of the eighteenth century and was characterised by the introduction of water and steam power to mechanise the way of doing things. The second industrial revolution began at the beginning of the twentieth century, this revolution focused on electric power for the mass production process (Kagermann et al. 2013). The third industrial revolution emphasised digitalisation by introducing microelectronics and automation, and it did so by automating manufacturing using electronics and information communication technologies (Rojko 2017). The fourth industrial revolution can be traced to Germany in early 2012. Since then, it has swept around Europe and Asia. A complete link between individuals and the workplace is now possible thanks to this change (Rojko 2017; Osunsanmi et al. 2018). This revolution has enabled a total connection between people and the work environment. It is characterised by the internet and the speed at which work is done and is customer friendly. Crnjac et al. (2017) asserted that the idea of smart factories represents the fourth industrial revolution; nevertheless, these smart factories employ a very different method of manufacturing.

Kagermann et al. (2013) posited that facility 4.0 is based on cyber-physical systems, automation, cloud computing, 3D printing, BIM, robotics, augmented reality, drones and big data. Schwab (2016) pointed to nine major systems and processes that constitute facility 4.0 which are the IoT, autonomous robots, big data and augmented reality. The fourth industrial revolution, according to Rojko (2017), is not just focused on the cyber-physical system (CPS), but also on the role played by the industrial IoT in the system of industrial production. In Africa, a study by Osunsanmi et al. (2018) showed that the South African manufacturing industry was able to adopt the fourth industrial revolution in such a way, it improved the industry's performance. Furthermore, the failure of the FM industry not to adopt and explore the benefits of digitalisation has significantly affected the FM industry. Additionally, because developing-country sectors have not undergone a considerable shift, the FM business has been negatively impacted by its reluctance to accept and explore the possibilities of digitalisation (Castagnino et al. 2016). The FM industry in Nigeria is facing a model shift as new technology is taking over the world but the FM industry's level of adoption is at a slow pace. Ezeokoli et al. (2016) noted that some industries in Nigeria have chosen to continue to dwell on paper-based practices and processes.

The majority of industries, including those in the banking sector, automotive industry, manufacturing industry, entertainment industry, and so on, have embraced the fourth industrial revolution to enhance consumer behaviour and the performance of the respective businesses (Osunsanmi et al. 2018). Regrettably, the FM industry has lagged because of the slow uptake of digital innovations. The development of the construction and FM industries has been hampered by the low acceptance and deployment of digital technology. This study admits that several published studies on the fourth industrial revolution and digitalisation have been conducted. For instance, a study on building 4.0 in South Africa was carried out by (Osunsanmi et al. 2018), (using a quantitative research approach). Aghimien et al. (2018) conducted a study on the digitalisation of the construction sector in South Africa, but from the perspective of a construction professional (using a quantitative research approach). In addition to studying the digital revolution in Nigeria, Ezeokoli et al. (2016) looked at it from the perspective of professionals (using a quantitative research approach). However, there isn't much information in the literature on Nigeria's FM industry's digitalisation and facility 4.0. This study adopts a qualitative research approach and contributes to the body of knowledge as it reveals the development surrounding facility 4.0 and digitalisation of the facility industry in Nigeria but from FM firms' perspective.

In addition, despite the advantages of digital transformation in the FM sector that have been outlined by several studies in the past, the FM industry in Nigeria has been hampered by important reasons. Several studies have investigated digital transformation, digitalisation, and industry 4.0 all over the world (Dale 2007; Ibem and Laryea 2014; Ernest and Young 2015; Gerbert et al. 2016; Berger 2016; Ezeokoli et al. 2016; Castagnino et al. 2016; Rouse 2017; CGI 2017; Aghimien et al. 2018; Osunsanmi et al. 2018; Ofoefule 2018; Alaloul et al. 2020; Asuquo et al. 2020; Ling et al. 2020; Neuman et al. 2021) just to mention a few. Despite the benefits of digital transformation, significant factors have hindered the digital transformation of the FM industry in Nigeria. There is a need to unravel these factors within Nigeria's FM industry. Several studies operationalise the factors inhibiting digital transformation based on the peculiarities of the country (Alaloul et al. 2020; Aghimien et al. 2018; Mukwawaya et al. 2018; Bhuiyan et al. 2020; Ling et al. 2020; Moshood et al. 2020). The reviewed studies centred on a questionnaire survey and literature review, providing a gap in the need for a qualitative research approach. The reviewed studies centred on a questionnaire survey and literature review, providing a gap in the need for a qualitative research approach. The study intends to establish the factors inhibiting the adoption of the fourth industrial revolution and the digitalisation of the FM industry in Nigeria. It also intends to proffer solutions through effective strategies that will change and improve the performance of the country's FM firms. The dearth of studies examining factors preventing their digital transformation is sketchy.

3.2 Review of Related Literature

3.2.1 The Concepts and Digitalisation of the FM Industry

Ernest and Young (2015) posited the idea of digitalisation as a procedure involving the conversion of analogue or conventional methods of information transmission into digital processes or information, which implies the intersection of new technology and contemporary skills in altering and enhancing customer behaviour. The act of organising and changing information into a clearly defined collection that is digital in identity is the idea of digitalisation (Rouse 2017). The idea of "digitalisation" has gained traction to boost output, increase operational effectiveness, and provide firms with more chances to succeed (Aghimien et al. 2018). Digital transformation is engineering knowledge that manages the formation of electronic devices, methods and systems to make communication more accurate and flexible (Lee et al. 2021). Big data, the internet of Things, digital twins and blockchain are digital technological tools in the real estate business (Bröchner et al. 2019; Lee et al. 2021).

The digitalisation of the FM industry offers a speed of reaction to change in the global marketplace, which will increase operational efficiency and in return ensure higher productivity (Ofoefule 2018), and there are four principles of digital transformation, these principles evolve around operations, products/services, employees and clients/customers. With customers, devices, and organisational or FM practices at the centre of change, the key advancement of digitalisation is to make FM firm operations digital. This will increase the agility, profitability, and overall performance of the FM firms, allowing them to compete for digital clients (Bahl 2015; EMC 2016). The degree of digital maturity is another important facet of digitalisation. The construction and real estate sectors, on the other hand, were found to have the lowest digital maturity levels in Africa. It was proposed that to stay competitive, FM enterprises and the industry adapt how they conduct business in the digital age (Ezeokoli et al. 2016; Oladapo 2007). Most stakeholders in several sectors nowadays are aware that digitalisation is increasingly changing every aspect of business (Berger 2016). Digitalisation will present FM firms with opportunities to enhance overall performance (Osunsanmi et al. 2018). Since the advent of BIM, digitalisation through digital technology has only lately been a widespread strategy among those involved in the built environment (Ibem and Laryea 2014). Aghimien et al. (2018) recommended that BIM be completely integrated across all phases of construction activities, particularly the operating phase.

The digitalisation of the FM industry entails providing services intelligently and shrewdly by assembling data using cutting-edge new technology gadgets. This will make it simple to analyse data and make decisions that will increase the FM industry's productivity and efficiency. With the FM sector going digital, businesses might be able to collect data, and sophisticated analytics can also boost productivity, timeliness, and risk assessment (Agarwal et al. 2016). To provide services with less effort, more accuracy, and increased productivity, FM businesses must integrate digital advances (Aghimien et al. 2018). Therefore, for the FM sector to remain competitive, digitalisation is a need. Berger (2016), digitalisation involves firms encountering connected systems at every stage of the supply chain. Since these digital tools and techniques are based on information and communication technology (ICT), digitalisation is the process of utilising them, and it will fundamentally alter how FM services are provided (Berger 2016). The low level of awareness among FM professionals is one of the biggest obstacles to the digitalisation of the FM industry. Information and communication technology adoption is not at a high level (Ezeokoli et al. 2016). Therefore, FM firms must go digital to remain competitive. Halim (2010) added no FM business or firm can effectively solve its growth difficulties unless its objectives and plans for accomplishing them are rooted in the field of knowledge production and powered by digital technology.

Digitalisation has revolutionized knowledgebase industries (Wong et al. 2018), improving access to information and communication speed (Lee et al. 2021). Mirarchi et al. (2018) proposed a framework for integrating digital technology into FM, allowing end users to integrate with maintenance processes. Digital transformation has enabled automation in the construction industry, and BIM has been used to capture, transfer, and store big data/information. A data management system for asset maintenance can have a significant effect on building performance (Lee et al. 2021).

The major goal of the Fourth Industrial Revolution (4IR) is the digitisation and automation of the Architectural, Engineering and Construction (AEC) as well as the

Facilities Management industry which among others involve the use of robotics and other technologies to deliver different phases in the life cycle of construction projects (García de Soto et al. 2022). One of the major digital technologies that have transformed the ways facilities are managed is the Internet of Things (IoTs). According to Oke et al. (2020), it is a network of interconnected devices that are embedded with network connectivity, sensors, software that enable them to amass and exchange data making them responsive. In other words, IoTs embedded in buildings and infrastructure automatically disconnects systems not in use thereby ensuring sustainability. Although, the global FM industry has witnessed unprecedented transformation as a result of the impact of technology, its impact on the FM industry of developing countries such as Nigeria is unclear. This study examines impediments to full adoption of digitisation in Nigeria's FM industry.

Studies from developed nations, Helfert et al. (2020) and Miehlbradt et al. (2021) examined the digitalization of Facility Management and identified gaps and opportunities for further research. They also highlighted the benefits and challenges of digitalisation and provided recommendations for practitioners. Egilmez et al. (2020) present a case study of a smart building project using digital technologies to optimize facility management operations. Fuchs et al. (2020) developed a comprehensive Framework for digitalization in FM, taking into account the needs and requirements of different stakeholders. These studies demonstrate the growing interest in the digitalisation of FM practices, which can improve efficiency, reduce costs, and provide better services to occupants. Industry reports and surveys provide insights into the current state of digitalisation in facility management, such as Deloitte (2021) and Verdantix (2020). These studies indicate that facility managers need to adopt a strategic and holistic approach to digitalization to address challenges such as interoperability, data privacy, and technology obsolescence.

Over the past several years, there has been an increase in interest in DT adoption in FM, and numerous governments (e.g. Australia and the UK) have emphasized the need to revolutionize the FM industry by accelerating DT adoption (Wong et al. 2018). Digital tools have been used to revolutionise the FM sector in emerging economies, but there is still a lack of research and implementation in this area. Recent studies have focused on potential applications in FM practice. Despite, the growing concerns about the digitalisation of FM practice across the globe, studies on facility 4.0 were non-empirical in nature (Wong et al. 2018; Mirarchi et al. 2018; Bröchner et al. 2019; Lee et al. 2021), empirical studies are missing in literatures. However, in developing nations like Nigeria, there are limited studies on facility 4.0. Which this study intends to explore the factors inhibiting the digital transformation of FM practices and proffer effective strategies for implementing facility 4.0 in developing nations.

3.2.2 *Standardisation of the FM Industry in Nigeria Through BIM*

The BIM application has gotten little attention in the Nigerian FM sector. Adebimpe and Moses (2016) suggested that professionals are not very aware of the use of BIM. Alufohai (2012) noted that despite the multiple advantages obtained by the several nations that have completely embraced BIM, awareness of and acceptance of it in Nigeria's business and public sectors as well as among diverse FM experts has been relatively gradual. To encourage the use of BIM in Nigeria and raise awareness of the method, the tools used, and the advantages of its adoption. Ibrahim and Birshir (2012) argued that investigations into the degree of understanding and usage of BIM in Nigeria are necessary. Therefore, there is a need to make BIM standards for all FM stakeholders.

According to a study by Berger (2016), the "2014 EU directive" recommends using BIM as one of the criteria for awarding public contracts. Additionally, the multi-phase "Digital Planning and Construction" plan, which was unveiled on December 15, 2015, by Germany's Federal Ministry of Transport and Digital Infrastructure, follows the directive's lead and makes using BIM mandatory for public infrastructure projects in Germany as of 2020. The BIM program was launched as a result of cooperative efforts between the government and the construction industry in the United Kingdom (UK). Its goal is to encourage the use of BIM by public and commercial organisations involved in the acquisition and delivery of infrastructures and buildings to decrease costs, improve operational efficiency, support more efficient and effective supply chains, and contribute to the development of a forward-thinking society (Wix and Nicholas 2008; Adebimpe and Moses 2016). The Netherlands, Denmark, Finland, and Norway already have similar laws in place (Berger 2016). The situation in Nigeria, however, differs from that in other countries, according to the research findings of Ibrahim and Birshir (2012), who noted that there is no government backing for the use of BIM in the construction project execution or FM practices, private clients are unaware of it, and the professional groups pertinent to the FM sector have not taken action to teach and retrain its members on its usage and application. In conclusion, Adebimpe and Moses (2016) advised the Nigerian government to create legislation mandating the use of BIM in all phases of building projects to improve quality, save costs, eradicate corruption, and uphold the law. By doing this, it would motivate the regulatory organizations responsible for regulating the operations of FM experts in Nigeria to make sure that BIM training and refresher courses are provided to all parties involved in the FM business.

3.2.3 *The Factors Inhibiting Adoption of Digital Transformation by FM Firms in Nigeria Under the Fourth Industrial Revolution*

Digital transformation of facility management practices involves the integration of digital technologies into the management of building systems and services in order to improve efficiency, reduce costs, and enhance the overall user experience. The adoption of digital technologies in facility management practices is influenced by organizational culture, leadership support, technology infrastructure, and user acceptance (Bougourd et al. 2020; Li et al. 2021). In this response, this study provided empirical research on these factors in the context of the adoption of digital transformation of facility management practices.

Fitzgerald et al. (2013) noted that there was no pressing need to embrace the digital transition. Mäkinen (2017) provided more evidence to support the claim that implementing digital transformation is hampered by a lack of urgency. This happens when there is a drop in knowledge of digital potential and motivation to work toward digital objectives, which leads to complacency (Fitzgerald et al. 2013). One of the main obstacles to the implementation of digital transformation is the capacity to do nothing or the capacity to foresee embracing digital trends. Lack of vision has hindered so many companies from going digital (Jadhav et al. 2019), while poor digital culture and low level of digitalisation have posed a threat to the transformation of industries in Nigeria. Ezeokoli et al. (2016) posited that causes include not developing a digital mindset and managing technological change. Both (Mäkinen 2017; Fitzgerald et al. 2013) claimed to a lack of vision inhibits the implementation of facility 4.0 by FM firms. The high cost of obtaining and using digital technologies as a result of their complexity and high price are additional concerns (Deloitte 2017; Oke et al. 2018). The competitive advantage of FM companies that don't invest will diminish (PWC 2016). Fitzgerald et al. (2013), Mäkinen (2017) suggested that another issue impeding the adoption of digital transformation is a lack of funding. Due to the significant degree of investment required for digital transformation (PWC 2016). Fitzgerald et al. (2013) stated further that a lack of resources, rather than only a lack of money, was what prevented the adoption of digital transformation under facility 4.0 (manpower).

External factors, such as regulatory requirements and market pressures, can influence the adoption of digital transformation of facility management practices (Bougourd et al. 2020). In addition, market pressures, such as competition and changing customer expectations, can drive organizations to adopt digital transformation practices to remain competitive (Li et al. 2021). Strategic planning can help organizations ensure their initiatives are aligned with their business goals and invest resources in the areas that will impact operations the most. However, the sophistication of digital technologies makes them hard to buy, operate, and work with and will demand specialist skills, making all procedures costly to acquire (Oke et al. 2018). Smaller FM firms, especially those in emerging nations, may have substantial difficulty understanding the concept of facility 4.0 since it is not yet widely adopted.

The adoption of facility 4.0 will need significant capital expenditure, at least initially (Vaidya et al. 2018). Facility 4.0 and the digitalisation of the FM sector will increase the need for workers in some fields. A study by Deloitte (2017) found that the FM business would become more digitalised, requiring new technical capabilities and that FM firms are struggling to find the requisite talent. Ezeokoli et al. (2016) also highlighted the issue of a lack of qualified specialists. In addition to the high expense of purchasing machines, Andrea (2019) noted that it might be difficult for a small business to locate trained people to handle and run these devices. The lack of employee incentives will also hinder the adoption of digital trends and present further hurdles for the current workforce (Ezeokoli et al. 2016).

Bhuiyan et al. (2020) opined that workforce prices make the FM industry unenthusiastic about automation, as it requires a large amount of investment. Nigeria's present labour force lack skills, competence, knowledge, and expertise in the utilisation of IoT, and data analytics among many others. Olaitan et al. (2021) further asserted that shortage of skilled workforce and technological infrastructure plays a key factor in going digital while lack of it poses a serious threat to digital transformation.

The workforce of the FM firm's lack of skills and competence is a barrier to the implementation of digital transformation, thus it's important to collaborate with other organisations to offer proper training on current digital technology developments (PWC 2016). This was corroborated by Mäkinen (2017), who noted that the most frequently cited barrier to the implementation of digital transformation is a lack of skills and competencies. The inadequate availability of the appropriate digital skills and capacities was another issue (Ezeokoli et al. 2016). Competent personnel with digital capabilities are required to enable digital transformation, but this process will be hampered by inadequate skill sets. An improved competence will enhance the possibilities of data analytics (PWC 2016). Oke et al. (2018) added that not only a lack of training but also the expense of digital training are factors preventing the implementation of digital transformation from a digital collaboration perspective. To get over these issues, specific training or retraining of current personnel in using new applications and technology would be needed (Deloitte 2017). Facility 4.0 and the digitalisation of the FM industry are being hampered by a lack of a digital mindset, a lack of a clear digital operations vision and top management support, a lack of digital culture and training, problems with interoperability, and poorly adaptable standards (Ezeokoli et al. 2016; PWC 2016).

According to Mäkinen (2017), the aspect of lack of teamwork will prevent organisations from successfully implementing digital initiatives as both business practices and products become more complicated. Ezeokoli et al. (2016) noted lack of cooperation hurts sharing culture and stunts the development of any industry. Balfour Beatty (2017) added that technologies like the internet of things (IoT) and virtual/augmented reality require greater resources and more data use, which poses a problem with rising energy demand. These will lead to the creation of new energy technologies and guarantee that resources are not irreparably exhausted. Osunsanmi et al. (2018) observed that FM business has yet to reap the rewards of the fourth industrial revolution, which may be because ICT adoption has been slow to take off. Osunsanmi et al. (2018)

further added that very few FM specialists are aware of the fourth industrial revolution (facility 4.0). Ezeokoli et al. (2016) noted further that the implementation of digital technologies was still in its infancy because of a lack of technical and critical understanding, among other management terms.

Data security is facing significant difficulties as a result of the internet of things, smart technology, and the interaction of data between the real world and the internet. Data security issues and worries about losing control of your company's intellectual property were highlighted (PWC 2016). Security (data security and privacy) is noted by Ezeokoli et al. (2016); Oke et al. (2018) as a problem of digital transformation. Balfour Beatty (2017) suggested that the use of smart and digital technologies will raise the dangers of cyberattacks, and these risks will persist since FM enterprises and the industry as a whole will rely on digital technology. Furthermore, Deloitte (2017) opined that the use of smart technologies like the internet of things (IoT), the internet of services, data, and people will foster new opportunities for data theft, hacker assaults (cybercrimes), and industrial espionage. Strange and Zucchella (2017) highlighted that the adoption of the internet of things (IoT) would increase concerns about cyber-security and hazards as a result of the possible entry points for hackers provided by the billions of sensing devices and communications devices. Consequently, if cyber-attacks are not recognised, they will undoubtedly have a disastrous effect on facility 4.0.

3.2.4 Strategies for Aiding Digital Transformation in the Nigeria FM Sector Under the Fourth Industrial Revolution

Some approaches can support Nigerian FM companies on their path to digital transformation so they may benefit from digitalisation's higher efficiency, improved communication, cost savings, and innovation. As a reaction to adopting facility 4.0, digital disruption of the FM business necessitates a digital cooperation strategy. Depending on an organisation's digital capabilities, the roadmap or plan that will support digital transformation under facility 4.0 will be developed (CGI 2017). For FM enterprises to adapt to facility 4.0, they must concentrate on and enhance their digital competencies. Deloitte survey (2017) established a strategy for incorporating facility 4.0 and digitalisation through digital collaboration. This technology is also making it simpler to establish effective alliances and partnerships with third parties who can provide various value chain components. Schwab (2016) noted that data-based services and asset performance through analytics necessitate a new kind of digital cooperation given the magnitude and rate of technological disruption as well as user experience. As a result, to supply services and enhance their performance, FM enterprises and sectors must develop partnerships and alliances.

Facility 4.0 is focused on the clients/customers, which is another approach FM firm must understand to apply it. In the digital age, the focus is on the consumer, as well as how they are provided (Schwab 2016). The quality of a firm's connection with its client is becoming ever more important, and clients/customers are growing knowledgeable and strong thanks to the information at their fingertips, thus FM businesses cannot depend and focus on one or few channels of information and influence (Deloitte 2017). FM firms must match the power of consumers, which will now shape the operations of the firms around the customers, as they have access to more data, options, and influence daily (Deloitte 2017). Schwab (2016) observed consumers' expectations have changed into an experience, and facility 4.0 has provided insights into customers' needs and behaviours that are used to inform marketing and sales strategies.

According to Agarwal et al. (2016), senior management must explain why these changes are necessary and what they represent for organisational structure, capabilities, and resource allocation to effectively implement change management as a strategy. As a result, organisations that don't engage in change management are more likely to encounter resistance and fail (Agarwal et al. 2016). Redefining the business model of FM enterprises and reinvesting in the future are two more strategies. FM firms must also run more cost-effectively and leanly (Deloitte 2017). In addition, Deloitte (2017) demonstrated how the implementation of automation and cognitive technology may result in cost savings that are significantly larger than historical standards, enabling FM organisations to pursue rapid expansion within the same period. The necessity for FM businesses, however, is to provide personnel (project participants) with the skillsets required to carry out digital technology operations. The training programs may be developed to instruct and educate FM professionals on the practical skills necessary for successful implementations of digital technologies within a certain discipline (Oke et al. 2018). The management of FM enterprises may fully benefit from this training to develop FM employees in the use of digital technologies to explore the advantages that come with digitalisation. The training can be organised and instructed by software vendors (experts) at the training facilities (Becerik 2004).

Oke et al. (2018) recommended solutions including fostering an environment for appropriate training and education on digital cooperation as well as enabling easy access to loans and assistance from the government, financial institutions, and businesses in the FM sector so that FM enterprises have the financial will to adopt facility 4.0 and the digitalisation of the FM industry in general. Three measures were proposed by Mutesi and Kyakula (2009) to support the digital transformation: government loans for technology advancement; upgrading and hiring experts with technological capabilities, and enough staff training for FM personnel. Kundishora (2006) suggested strengthening digital technology and fostering collaboration require actions like government loans, continuous electricity supply, lowering the cost of digital tools like computers, ensuring that every department creates and manages a computerised data system, spreading awareness of the use of technology, and supporting local research and development. Oke et al. (2018) added the need for increased research and development as it will help with the digital revolution.

However, undergoing digital transformation under the facility 4.0, high level of digital competence is required. Himang et al. (2020) posited that it is vital for the competence level of 4IR adopters to be high in the aspect of installation, maintenance, security, operation, and network administration. Therefore, technical support and expertise are key for implementing 4IR. There is a need for construction firms to train workers with the necessary abilities to carry out digital technology activities.

Any nation that wants to genuinely develop and prosper must go beyond digital skills and harness its research and development capability. The involvement of firms in technological training is poor. 4IR skills depend on the extent and quality of human resources and the current use of digital technology and platforms. The construction industry in Nigeria has been widely criticized for depending on manual labour, in implementing 4IR construction firms need the right skills to digitalise their operation. Olufemi (2020) observed that 4IR implementation can only be as energetic as the level of digital literacy of the firm. Dall'Omo (2017) in a report on digitalisation maturity in Africa revealed that Nigeria has a low level of maturity when compared to South Africa.

According to World Economic Forum (2020), not just Nigeria but Africa as a whole have a digital skills gap and by 2030 Nigeria will need digital skills for about 35–45% of its job. These digital skills vary from digital literacy, amination, software testing, technical support, data storage technologies, human–computer interaction, data sciences, cybersecurity, and artificial intelligence among many others. Furthermore, strategies are effective in the adoption of 4IR, as they serve as a medium to implement 4IR components. According to a report by the World Economic Forum, developing a digital strategy is essential for successful digitalization in facility management practices (World Economic Forum 2019).

According to Eze et al. (2015), the government must assist FM enterprises in Nigeria, especially to increase their finance and financing capabilities. The study by Oke et al. (2018) noticed the approach of individual training of personnel by businesses, and the study further proposed that digital technologies should be integrated and taught as a module in higher institutions to generate graduates who are oriented toward digital technology. This means students are exposed to BIM applications. Adebimpe and Moses (2016) urge the Nigerian government to pass legislation mandating the use of BIM in building projects to improve quality, save costs, combat corruption, and uphold the rule of law. BIM should be completely integrated across all stages of building operations (Aghimien et al. 2018). There is a need to encourage the implementation of facility 4.0 in the FM industry by conducting a scientific investigation on effective strategies as it was discovered from the review of the strategies for assisting digital transformation under the fourth industrial revolution that only a few studies had been conducted in Nigeria. Based on this assertion, the study aims to add to the knowledge base in these fields, which serves as the basis for the study.

The study, which grew out of the disruptive innovation theory, made clear that the connection between digitalisation and performance is based on how digital tools transform the current construction market by presenting ease, accessibility, suitability, and affordability where barriers and high costs are the status quo (Christensen and Raynor 2003). Innovative performance and market expansion through disruptive innovation have the potential to disrupt current market relationships. The digitisation of the FM practices, otherwise known as facility 4.0 has become a disruptive wave of change that affects the construction industry to the extent that one of the most critical elements is the implementation of the technological revolution (Lekan et al. 2020; Mansour et al. 2021). Digital tools such as BIM, drones, the internet of things, 3D printing, and robots among many others are digitalisation that disrupts the current market. Based on this theory, this study seeks to establish factors affecting the digital disruption of FM practices and proffer effective strategies for the implementation of facility 4.0.

3.3 Methodology

The interpretivism method serves as the study's philosophical foundation (epistemology assumptions). The interpretivism method examines the complexity, richness, and depth of the relevant phenomena (Wijesinghe 2011). The interpretivism method, which forms the foundation of qualitative research, enables the researcher to comprehend social phenomena from the perspective of human contexts (Saunders et al. 2009). To develop this strategy, the study used a case study research methodology, and FM firms operating in Lagos, Nigeria, provided qualitative data. When the borders between the phenomena and the context are not immediately obvious, Yin (2003) argued that a case study is an empirical investigation that looks into the real-world context of a current phenomenon. To achieve the study's goals, which include identifying the barriers to digital transformation adoption and formulating plans to support it among Nigerian FM enterprises in the context of facility 4.0. Because the study used a case study methodology, it is appropriate to use a purposive sampling technique. Previous studies adopted this method to provide deeper insight on digitalisation (e.g. Egilmez et al. 2020; Oladokun et al. 2021). When a researcher intends to create a historical reality, explain a phenomenon, or create something about which little is known, purposeful sampling is very helpful. In qualitative research, the purposive sampling approach is more frequently employed (Kumar 2011). Since FM managers supervise the operation of the case businesses, the research considered the information they gave was enough. In their absence, any stakeholder with a sufficient understanding of ICT or technical innovation was interviewed, and this also offered a good quantity of data for analysis. The interview served as the primary data-gathering tool for the study. The interview is by far the most often employed strategy for conducting or gathering qualitative design research, and it was utilised in this study (Kumar 2011). The interview was used to get a thorough grasp of facility 4.0 and digitalisation as well as to comprehend underlying causes, viewpoints, and motives to find thought patterns and further our understanding of facility 4.0 (Oladokun et al. 2021). In a group of four international and four domestic construction enterprises, eight companies were questioned. Multiple case studies are a feature of the study. Eisenhardt (1989) recommended four to ten instances for multiple case studies, arguing that less than four cases would make it difficult to generalise since there isn't much complexity, while more than ten cases might make the sophistication and volume too overwhelming. The research used theme analysis to examine the qualitative data that was produced. Thematic analysis is the most used type of qualitative analysis, according to Guest (2012). The specific approach used was conversational analysis, which is used to identify, examine, and document patterns of significance or themes within data. A thorough investigation of the discussion between two or more persons is the subject of conversation analysis. In qualitative research, data analysis is used to recognise or locate parallels, discrepancies, patterns, and descriptions of the subject under investigation. Oladokun et al. (2021) utilized thematic analysis to deduce discrepancies, patterns, and descriptions on benefits of going digital.

3.4 Data Presentation, Analysis and Discussion of Findings

3.4.1 Specifics of 'Case' Firms Archetypal

Table 3.1 designates details of the case firms' representatives, including the position held within the company, the representative's profession, and their individual work history within the FM sector as a whole.

Table 3.1 indicates the categorisation of the case firms into foreign and indigenous FM firms (four foreign and four indigenous, in a total of eight), which were interviewed to have an in-depth understanding of critical context-based issues fostering low adoption of the digitalisation transformation of FM firms in Nigeria. Table 3.1

Table 3.1 Specifics of 'case' firms archetypal

S/N	Company type	Categorisation of the company	Position held	Profession	Working experience
1	FM firm A	Foreign	Facility manager	FM	12 years in FM practice
2	FM firm B	Foreign	Facility manager	FM	21 years in FM business
3	FM firm C	Foreign	Facility manager	FM	10 years in FM business
4	FM firm D	Foreign	FM specialist	FM	20 years in the FM/ construction business
5	FM firm E	Indigenous	FM consultant	Architecture	10 years in the real estate business
6	FM firm F	Indigenous	FM specialist	Quantity surveying	15 years in FM business
7	FM firm G	Indigenous	CEO	FM	18 years in business
8	FM firm H	Indigenous	Facility manager	FM	11 years in FM business

Source Author's Field data (2022)

also indicates the position held by the case firms' representatives. Four representatives are facility managers (firms A, B, C and H). FM firm's 'D' and 'F' representatives are FM specialists, while FM firm E is an FM consultant. FM firm G was represented by the CEO. Table 3.1 further indicates the profession of the case firms' representatives. FM firms A, B, C, D, G and H, representatives are registered FM professionals (Association of facilities management practitioners in Nigeria), while FM firm F representative is a registered member of the quantity surveyor registration board in Nigeria, and E representative is a registered architect (Nigerian institute of architects). Table 3.1 further the working experience in the respective case firms and the FM industry at large. FM firm A had 12 years of working experience in the FM industry, FM firm B had 21 years, and firm D had 20 years of working experience both in FM and construction industries. FM firms C and E had 10 years of working experience in the FM industry, and firms F, G and H had 15 years, 18 years and 11 years of working experience, respectively. The result of the respondent's characteristics is an indication that they are reasonably knowledgeable as to depend on their opinion.

3.4.2 Factors Inhibiting the Adoption of Digital Transformation Under the Fourth Industrial Revolution

There are so many things that prevent the fourth industrial revolution's acceptance of digital transformation. The case firms surveyed throughout the industrial revolution indicated several things, including a lack of resources and knowledge and a paucity of people with the necessary training (Table 3.2).

Table 3.2 reveals that seven out of eight case businesses identified a lack of finances (cost) as the barrier preventing their FM enterprises from adopting digital transformation. To be fair and honest, it costs money to implement the fourth industrial revolution; this is stated by FM business G'. "These advances, and ought I say digital revolution, capital vast," the FM business D' noted. Additionally, FM firm C' stated, "These things are expensive and need significant funding to accomplish because they are not done regularly, thus we must spend additional funds to do so." The expense of putting certain things in place is significant, and a lack of motivation is a contributing problem, according to FM business B'.

The initial cost is one of the concerns, according to FM firm "H," which noted that "you can't deliver what you don't have" and that "cost is number one" when it comes to implementing these kinds of systems. While FM businesses B, E, F, and H' concurred that a barrier is a lack of knowledge (Table 3.2). FM company B' continued by saying that while a lack of understanding is a role, the infrastructure that most of these things rely on is lacking. Factors including ignorance, indifference, and a lack of an enabling atmosphere were also identified by FM firm H. While FM firm E' concurred, reasons including insufficient knowledge, managing digital technologies,

Table 3.2 Major findings on the factors inhibiting adoption of digital transformation under the fourth industrial revolution

Key Info. provider	Major interview findings	Issues address
FM firm A	"The company only runs into one obstacle to adopting digital transformation"	Low technical knowledge was the factor that emerged
FM firm B	"There are two to three considerations for this firm"	Lack of funding, inadequate resources, and a lack of the necessary infrastructure were the reasons that emerged
FM firm C	"The company first encountered a few obstacles that prevented the implementation of digital transformation"	Lack of funding, a lack of resources, and the complexity of these technologies were the issues identified
FM firm D	"Yes, there are barriers to the adoption of all technology, and this company was no exception"	The causes identified were a lack of resources and a paucity of skilled workers
FM firm E	"There are certain initial barriers to adopting digital transformation"	Limited technical expertise, low awareness, a lack of resources, and the cost and complexity of technology were the reasons that emerged
FM firm F	"Honestly, there are a few reasons why the company is seeing low uptake"	The issues included poor awareness, low technical expertise, low adoption, a lack of teamwork, a lack of resources, and a paucity of personnel with the necessary training
FM firm G	"Due to these challenges, the company is experiencing limited acceptance of digital transformation"	Lack of resources/funds was the factor that emerged (cost of acquiring)
FM firm H	"These issues, which prevent the adoption of digital transformation, are the cause of the company's poor adoption"	Lack of exposure, lack of knowledge, and a lack of an enabling environment were the causes identified

Source Author's Field Survey (2022)

a lack of resources, and the complexity and cost of digital tools were important. FM firm F' claimed,

> The lack of collaboration and the absence of the necessary skills and capabilities are other factors that affect us, as well as the FM industry. "Generally, one thing that affects us, not just us but the FM industry at large, is the low level of awareness of trends or events.

FM firm 'C' opined that some firms lack vision in adopting digital transformation. FM firm C' claimed these technologies are complex, they are not as easy to implement, "For instance, the RFID has a lot to do with barcoding, and all those stuff, the drones, you need systems, you need laptops, you need servers that will power the drones and make them useful". Another factor agreed was the shortage of trained professionals as FM firms F, D, and E' both agreed shortage of trained professionals is a factor inhibiting the adoption of digital transformation (Table 3.2). However, technical know-how is another factor revealed and was agreed upon by FM firms A, E, and F. FM firm A' added,

> I can say that one of the reasons we have so many foreign ex-pats here is because of the shortage of technical know-how, particularly among local employees. This also brought up the topic of training and retraining periodically so that people would be able to keep up with these innovations.

The difficulties that the international and indigenous FM enterprises handled were comparable and different, as seen in Table 3.3. Lack of resources, ignorance, a lack of qualified personnel, a lack of technical expertise, the high cost and complexity of these technologies, and similar factors apply to both international and domestic FM enterprises (Table 3.3). The main distinction between foreign and indigenous FM firms regarding the factors impeding digital transformation is that, compared to foreign FM firms, indigenous FM firms are more affected by the factors revealed (lack of funds, lack of awareness, shortage of trained professionals, and low technical know-how).

Table 3.3 Cross-case analysis of interview findings on factors

S/N	Similarities of issues address	Differences of issues address
1	Lack of money affected both foreigners and indigenous FM firms equally (FM firms B, C, D, E, F, G, and H)	When compared to international FM enterprises, indigenous FM firms indicated more about the lack of money as a barrier to digital transformation
2	Lack of awareness is common to both foreign and indigenous FM firms (FM firms B, E, F, and H)	When compared to international FM enterprises, indigenous FM firms reported more about a lack of understanding/awareness as a barrier to digital transformation
3	A shortage of trained professionals was revealed by both foreign and indigenous FM firms (FM firms D, E, and F)	When opposed to foreign FM enterprises, indigenous FM firms faced challenges such as a dearth of educated specialists
4	Low technical know-how was common to both foreign and indigenous FM firms (FM firms A, E, and F)	Comparatively speaking, domestic FM enterprises than overseas FM firms disclosed more about low technical know-how as a barrier to digital transformation
5	The expensiveness and complexity of these technologies was revealed by both foreign and indigenous FM firms (FM firm C and E)	There was no distinction between FM enterprises that were indigenous and foreign

Source Author's Field Survey (2022)

3.4.2.1 How Have These Factors Affected the Digital Operation of Your Firm?

In response to this question, FM firm H' suggested how the factors have affected the firm,

> For instance, I said the initial cost. Let me explain that since there isn't any money and the banks aren't pushing you to spend since they only consider the short term, it's challenging to get a financial institution to invest in your business for a 20 to 25-year period. This has an impact on the company's output and subsequently reduces the level of competitiveness with other ex-pats.

FM company C' said, "You need to start training when we need to implement these systems and procedures, or else these technologies will not be used properly." Because it takes time to solve the scarcity of qualified personnel, it has an impact on how the business operates. FM firm E' explained how the factors affected its company,

> Having the correct item in place at the right time takes time, so you have to send your personnel to learn and get trained. "We spent hours trying to find out what we could do, so at one point we sat down and looked for a method to combat the factors, so it limits the digital operation of the organization.

However, FM firm F' gave a straightforward response on how the factors have affected the firm,

> We are unable to complete projects on time, and because of our sluggish operations, we must completely embrace the fourth industrial revolution trend or risk losing the market.

"Lack of technological know-how has harmed our digital operation because when you have these gadgets you cannot utilize it, then it is of no use," FM firm A' further explained how the element had an impact on the business. According to FM business D', completing projects on schedule and competing were big issues. While FM firm B' said that the absence of expatriates is a constraint, FM firm G' stated that "Sometimes you may even lose clients."

3.4.2.2 What Have Been the Critical Factor and How Has Your Company Reacted to Them

Government policy (support), in the sense that there is not enough room for competition, is allegedly the main difficulty facing not only the corporation but the FM sector as a whole. Government organisations prefer to work with foreign corporations over domestic ones, thus even though it would have been better for the latter to be expanding or developing, it appears that the former was not given preference. Payment delays have also driven numerous FM firms out of business. Therefore, finding funding for a project from either private or public sources is difficult. The case companies responded by bringing in qualified experts. For FM firms D and F, the important reason was a scarcity of trained personnel. FM business F' responded further by planning several employee training and retraining seminars and workshops. The low technological know-how was identified as the key factor by FM businesses A, C, and D. The identical response was given by both parties: FM company C' stated, "We needed to bring in foreign experience, by which I mean we brought in our team in Europe". FM business A' said, "That's why, you find out we still have a lot of these foreigners in the company because the channel more of these innovations, the usage, and training sometimes."

Similar to FM firm E', which asserted that a lack of trained professionals and low technical know-how were critical factors, the firm responded by sending staff members abroad to study and research this innovation. The firm also invested in research and development and brought in foreign expatriates to assist with staff training and retraining. The main obstacle for FM businesses G and H was the expense. "Cost will come in, and we find a method to balance it," FM business G' continued. "If you want to stay relevant, you have to struggle to always secure those projects that will boost turnover yearly," FM company H' further stated. In conclusion, the relevant example firm identified several barriers to adopting digital transformation during the fourth industrial revolution. The most frequently discussed issues, albeit how the case companies are impacted by these digital transitions vary, include a lack of resources, poor awareness, a dearth of experts with the necessary training, a lack of technical know-how, and their expense and complexity.

3.4.3 Strategies for Aiding Digital Transformation Under the Fourth Industrial Revolution

Organisations and enterprises need to embrace enhancements, strategies, or methods to facilitate the adoption process to adopt digital transformation under facility 4.0. Effective tactics that FM firms should employ to promote the adoption of the fourth industrial revolution are shown in Table 3.4.

Table 3.4 showed the representative of the example firm's response to the fourth industrial revolution's tactics for assisting digital transformation. FM business E' offered the following set of solutions in response:

1. The client and how we service or satisfy their demands are our firm's main priorities. According to FM businesses C and F, clients are becoming more knowledgeable every day. A tactic is to concentrate on the client.
2. Digital collaboration: To share this expertise with us, we enlist the aid of other specialists, some of whom are still collaborating with us in person. Digital cooperation is an efficient technique for digital transformation, according to FM firm A-representative H's for the example firms.

Training and retraining were offered as techniques by FM businesses A, B, C, E, and F to support any digital collaboration (Table 3.4). "Training, we teach more individuals to get on board, it is not enough to have few people on board," FM business C' disclosed. FM company B continued, "I believe that first and foremost,

Table 3.4 Major findings on strategies of aiding digital transformation under the fourth industrial revolution

Key Info. provider	Major interview findings	Issues address
FM firm A	"Yes, there are tactics that help with digital transformation adoption"	The initiatives identified included government engagement and assistance, increased research and development, teaching about digital transformation in higher institutions, digital cooperation, training, and retraining
FM firm B	"Honestly, the company used certain strategies to aid digital transformation"	The revealed tactics included government backing, digital collaboration, and training
FM firm C	"The company needed these strategies to aid the adoption of digital transformation"	Government backing, more R&D, training, digital cooperation, and standardising BIM were among the tactics identified
FM firm D	"To encourage the adoption of digital transformation, methods must be used"	The tactics exposed included government backing, financing availability, an increase in R&D, digital cooperation, and standardising BIM
FM firm E	"Honestly without strategies, the company cannot adopt facility 4.0"	The tactics exposed included a client-centred approach, digital cooperation, training and retraining, standardising BIM, government assistance, and an increase in research and development
FM firm F	"These approaches assist in implementing the digital transition"	Government assistance, training and retraining, digital cooperation, teaching digital transformation in higher institutions, and an increase in R&D are among the methods identified
FM firm G	"We must find a means to get beyond the obstacles to the adoption of digital transformation"	The tactics exposed included training, government funding, adopting a digital mentality, and stepping up R&D
FM firm H	There are approaches to embracing facility 4.0, but the business is still researching some of them at the moment	The tactics were government sponsorship and digital collaboration

Source Author's Field Survey (2022)

the pertinent professional organisations should constantly refresh their members through seminars, conferences, and training." This was by FM firm H's assertion that "We have a federation of FM industry here in Nigeria, if definitely, they are aware of this sort of information at their conferences, they can as well present this as a major point to all FM businesses registered under them." Bring this technology to their doorsteps and explain how they can get started regardless of the level at which they operate; this would raise awareness.

Training and retraining, research, and development are listed as efficient solutions in Table 3.4. Additionally, the solutions recommended by FM company G, such as training and, if possible, government backing, will be helpful. The FM firms A, C, D, F, and F' promoted participation in or a boost in R&D. According to FM firm E', the business leads the industry in terms of investment in R&D. FM firm H' gave an honest response on research and development,

> You hardly see research and development in any construction firm. The only thing, I'll tell you, is that sometimes you hear about construction training their staffs is just about holding two or three days workshops or going on a conference or a retreat, but making that opportunity available whereby you will be to say let's send some people to go and learn outside the country, bring that information back to us display it, let's see how best we can adopt it, you hardly see that ha, you hardly see that here.

"I shall advocate government engagement in this fourth industrial revolution," FM company A' said. Digital transformation should be taught in academic institutions, according to FM businesses A, C, and F. FM company A' stated, "We need to evaluate our curricula to keep up with the reality of this transformation". BIM should be a standard, according to FM businesses E and C. Making BIM a standard and providing access to financing are other ideas that FM firm D' added, and they both, in my opinion, will help with digital transformation throughout the fourth industrial revolution. In conclusion, each example firm's representative provided a sound plan for completely implementing digital transformation in the context of the fourth industrial revolution.

The similarities and differences between the international and native FM enterprises were shown in Table 3.5. Both international and domestic FM enterprises frequently use tactics including digital cooperation, government backing, training and retraining, involvement/increase in research and development, making BIM a standard, and teaching digital transformation in tertiary institutions. Foreign and domestic FM enterprises collaborate digitally in the same way, and government assistance is equally evident. Compared to domestic FM enterprises, international FM firms, as shown in Table 3.5, demonstrated more successful methods.

3.4.3.1 How Can Government Support and Access to Loans Aid the Digital Transformation of Your Construction Firm?

How the topic matter may be done is the follow-up query, which aims to go further. A' argued that government assistance might not just come in the form of loans,

> I will say there are a lot of people in the country that one to go into business but you discover that sometimes, there have not been enabling environment for them and these are the area the government has to come in (areas of government policy making and creating enabling environment).

Government funding may be made accessible, according to FM firm H'. "Let's simply take this as an example, farmers are getting subsidies on equipment and fertilizers, in the same manner, if this is being provided to the FM firm," it was stated. FM business D' said that government assistance in the form of easier access to financing and favourable policy will provide the company with the assistance

Table 3.5 Cross-case analysis of interview findings of the strategies of aiding digital transformation under the fourth industrial revolution

S/N	Similarities of issues address	Differences of the issues address
1	Digital collaboration was common to both foreign and indigenous FM firms (FM firm A–H)	Regarding digital collaboration, there was no difference between the FM enterprises that were indigenous and foreign
2	Government support was common to both foreign and indigenous FM firms (FM firm A–H)	Regarding government backing, there were no disparities between the foreign and indigenous FM enterprises
3	Training and retraining were revealed by both foreign and indigenous FM firms (FM firms A, B, C, E, and F)	When compared to indigenous FM enterprises, foreign FM firms revealed more about training and retraining
4	Involvement/increase in research and development (R&D) were common to both foreign and indigenous FM firms (FM firms A, C, D, E, and H)	When compared to indigenous FM enterprises, foreign FM firms exhibited a greater involvement in or increase in R&D
5	Making BIM a standard was common to both foreign and indigenous FM firms (FM firms C, D, and F)	When compared to indigenous FM enterprises, foreign FM firms disclosed more about establishing BIM as a standard
6	Digital transformation should be taught in the tertiary institution was common to both foreign and indigenous FM firms (FM firms A, C, and F)	In comparison to indigenous FM enterprises, foreign FM firms disclosed more about the need for digital transformation education in higher education

Source Author's Field Survey (2022)

needed to fully embrace these trends. FM businesses F and B, however, asserted that the "government would not give you the loan" and that "even financial institutions are afraid of granting hefty loans, which are not what we need but if they can cut the interest rate it will assist."

Government support for FM company B's additions included favouritism of the local business, employment opportunities, and prompt payment. Payment from the government on time would be ideal since it would encourage the contractor or business to work with them. FM firm E' claimed,

> Government should patronise our company, give us jobs, provide the policy that suits the growth of indigenous firms, there should not be a delay in payment and government have to create an enabling environment for us to strive.

FM firm C' also gave similar thoughts, "basically the government has to do is to put in place for us, all that will be required of them, just structures and encourage the use of digital transformation, one thing I know it is important if digital transformation is integrated into the contract". FM firm G' claimed that government can support,

> Government backing would be very helpful and let me give you an example: the Chinese government sponsors their firms and provides them with all the instruments. Assuming we can obtain financing, it will help greatly.

The fourth industrial revolution's adoption of digital transformation might be facilitated by several forms of government support. The improvement of the FM sector will result in a greater uptake of digital innovations as a result of the assistance provided by favourable legislation, an enabling environment, a reduction in interest rates, fast payment, and the encouragement of digital transformation. In conclusion, the relevant example businesses presented a variety of aids for the fourth industrial revolution's digital transition. Although the methods used by the example firms to determine the tactics differ, government funding, digital cooperation, training, and increased research and development are the most often discussed ones.

3.5 Discussion of Findings

Each of the relevant ideas' classifications is displayed in Table 3.6. The patterns that emerged from each significant result and the cross-case examination of the interview data for each research aim are known as ideas. The analysis of the results is based on Table 3.6.

3.5.1 Factors Inhibiting the Adoption of Digital Transformation Under the Fourth Industrial Revolution

The results of this study also identified the barriers to the fourth industrial revolution's adoption of digital transformation. These issues include a lack of resources, a paucity of experts with the necessary training, a lack of technical expertise, a lack of incentives, a loss of enthusiasm, a lack of foresight, a lack of exposure, a lack of an enabling environment, and the cost and scope of digital technology. This result supports what Ezeokoli et al. (2016) found to be factors impeding digital transformation, including a lack of trained professionals, a lack of incentives, a lack of funding, a lack of the necessary skills and capabilities, a low level of adoption, and a lack of

Table 3.6 Categorisation of relevant concepts

S/N	Objectives of the study	Pattern
1	Factors inhibiting the adoption of digital transformation under the fourth industrial revolution	Lack of funds, lack/low awareness, shortage of trained professionals, low technical know-how, and expensiveness and complexity of these technologies
2	Strategies for aiding digital transformation under the fourth industrial revolution	Digital collaboration, government support, training and retraining, involvement/increase in research and development, making BIM a standard, and digital transformation should be taught in tertiary institution

Source Author's Field Survey (2022)

dynamism among managerial teams. Lack of vision has hindered so many companies from going digital (Jadhav et al. 2019), while poor digital culture and low level of digitalisation have posed a threat to the transformation of industries in Nigeria.

The study's findings are also consistent with those made by Osunsanmi et al. (2018), who suggested that FM stakeholders had little knowledge of the fourth industrial revolution. This study's findings support those of earlier research (Mäkinen 2017; Fitzgerald et al. 2013), which identified a lack of resources and a lack of vision as the two main factors/barriers impeding digital transformation. Bhuiyan et al. (2020) opined that digital transformation of FM industry requires a large amount of investment. According to Vaidya et al. (2018), facility 4.0 will necessitate a sizable initial capital investment, but it is still not widely adopted by SMEs. Smaller manufacturing companies, especially those situated in emerging economies, will have significant difficulty understanding the concept of the fourth industrial revolution as a whole. The main conclusion of this study is the high expense of obtaining the systems and procedures that made up the fourth industrial revolution, but it also shows how difficult it is to locate the trained personnel needed to handle and run facility 4.0.

This study's finding that the complexity and price of digital technology are deterrents to digital transformation is in line with that of Oke et al. (2018), who found the high cost of acquiring these digital technologies as a result of their high cost and complexity. This is because new technologies are pricey and complicated, making them difficult to run and expensive to purchase. They also require specialized training. Olaitan et al. (2021) further asserted that not just the high cost of acquiring digital tool is a challenge, but shortage of skilled workforce and technological infrastructure plays a key factor in going digital while lack of it poses a serious threat to digital transformation.

The study's conclusions showed that one element impeding digital transformation during the fourth industrial revolution was a lack of training. This was similar to PWC's (is similar to PWC's (2016) suggestion that there is a need to collaborate with outside organizations just to provide appropriate training on digital technological trends would be considered necessary or retraining current employees on using new applications or technologies, which will facilitate the adoption of the fourth industrial revolution. In conclusion, the study's key findings addressing the factors preventing digital transformation include a lack of funding, a lack of awareness, a paucity of people with the necessary training, and a lack of technical know-how across FM enterprises in Nigeria.

3.5.2 Strategies for Aiding Digital Transformation Under the Fourth Industrial Revolution

The results of this study showed that the Nigerian construction sector agrees that government help is a successful tactic. Government assistance for the sector in the form of legislation, financing, and enabling environments. This conclusion is consistent with Mustesi and Kyakula's (2009) recommendations that the government assist technical growth by offering to finance, enhancing and employing people with technological abilities, and giving FM employees/workers adequate staff training. Furthermore, FM enterprises in Nigeria should receive help from the government, according to Eze et al. (2015), so they may increase their fundraising and financing capabilities. The fourth industrial revolution must be implemented in the construction sector, which requires a digital collaboration strategy. Another important conclusion of this study is that digital collaboration is an excellent technique for assisting both international and domestic FM businesses. This conclusion is consistent with Deloitte's (2017) observation of digital collaborative efforts as a strategy, and this technology facilitates the formation of seamless alliances and collaborations with third parties that may provide various value chain components.

The study's findings showed that increasing R&D was an efficient method for promoting digital transformation. This conclusion is consistent with that of Oke et al. (2018), who found that increasing R&D will help the FM sector undergo a digital revolution. The study's conclusions showed that one technique for promoting digital transformation is periodic training and retraining of employees. According to this research, training programs should be developed to teach and inform FM workers about the necessary skills for using digital technologies effectively within the FM discipline. According to a report by the Royal Institution of Chartered Surveyors (RICS), training is essential to ensure facility managers and other stakeholders are proficient in using new digital technologies (RICS 2020). This training will raise FM workers' understanding of the fourth industrial revolution (Oke et al. 2018). Key area of the training should target the competence level in the usage of digital tools. Himang et al. (2020) posited that it is vital for the competence level of 4IR adopters to be high in the aspect of installation, maintenance, security, operation, and network administration. Therefore, technical support and expertise are key for implementing 4IR. These training could also be targeted on developing digital skills. These digital skills vary from digital literacy, amination, software testing, technical support, data storage technologies, human–computer interaction, data sciences, cybersecurity, and artificial intelligence among many others (World Economic Forum 2020).

The results of this study demonstrate that standardizing the FM sector using BIM is a useful tactic for promoting digital transformation. The adoption of BIM is essential because it will pave the way for additional future digital innovations in FM practice. This result is consistent with the remark made by Adebimpe and Moses (2016) that the Nigerian government, through its legislative branch, passed rules mandating the use of BIM in building projects to improve quality, save costs, get rid of corruption, and follow the law.

3.6 Conclusion and Recommendation

Since the advent of digitalisation during the fourth industrial revolution, the FM has faced serious challenges due to poor acceptance and low awareness. Although the effects of digitalisation may be seen in the final products of many industries, they are particularly noticeable in emerging countries across the world. Nigeria is one example of a developing country that has lagged in adapting to digitalisation. The fourth industrial revolution's digitalisation presents several issues for the FM in Nigeria, which the sector must analyse and devise solutions for. The major finding of the study shows that lack of funds, lack/low awareness, shortage of trained professionals, low technical know-how, and expensiveness and complexity of these technologies are factors inhibiting digital transformation under fourth industrial revolution. Strategies such as digital collaboration, government support, training and retraining, involvement/ increase in research and development, making BIM a standard, and digital transformation should be taught in tertiary institution are effective in the implementation of digital transformation of FM practices under the fourth industrial revolution. The study concluded that there are several reasons preventing the construction sector in Nigeria from adopting digital transformation as part of the fourth industrial revolution. But the most often cited barriers to adopting digital transformation during the fourth industrial revolution were lack of resources (the expense of acquiring digital technologies), poor awareness, a lack of qualified personnel, and a lack of technical know-how.

The digital transformation of FM firms may be reinforced by addressing the key issues impeding the adoption and use of digital tools throughout the strata of FM enterprises. Based on these findings, the study pushed the idea that Nigeria's FM sector has to go digital for any FM business to survive, and that the reasons it uncovered may vary across FM companies there. According to the study's findings, the operation and maintenance phase of a facility may perform poorly and have low returns worsened if FM firms do not prioritise digital transformation in the context of the fourth industrial revolution. This is because digital transformation transforms important phases of construction projects. In addition, government support in the form of favourable regulations, a supportive environment, digital cooperation, training, and increased research and development are key initiatives for the effective adoption of the fourth industrial revolution. Additionally, it may be said that the example companies' stated methods are successful, while they may not be the same as those of other FM firms. The adoption of the fourth industrial revolution and the digitalization of the FM sector in Nigeria might be led by FM enterprises and the whole FM stakeholder community. Collaboration between government and FM stakeholders is necessary for the implementation of the fourth industrial revolution. The fourth industrial revolution's adoption in Nigeria would elevate the FM sector to the top. The study recommends government officials and stakeholders in the FM sector work together to develop a suitable roadmap that would put the FM sector on the correct path for Nigeria's implementation of the fourth industrial revolution and digitalisation.

References

Adebimpe AO, Moses E (2016) Assessment of the level of awareness and limitations on the use of building information modeling in Lagos State. Int J Sci Res Publ 6(2):2250–3153

Agarwal R, Chandrasekaran S, Sridhar M (2016) The digital future of construction. Global Infrastructure Initiative. https://www.globalinfrastructureinitiative.com. Retrieved in Jan 2019

Aghimien D, Aigbavboa C, Oke A, Koloko N (2018) Digitalisation in construction industry: construction professionals perspective. Retrieved from https://www.researchgate.net/publication/329141252_DIGITALISATION_IN_CONSTRUCTION_INDUSTRY_CONSTRUCTION_PROFESSIONALS_PERSPECTIVE on Feb 2018

Alaloul WS, Liew MS, Zawawi AA, Kennedy IB (2020) Industrial revolution 4.0 in the construction industry: challenges and opportunities for stakeholders. Eng J 11:225–323

Alufohai J (2012) Adoption of building information modelling and Nigeria's quest for project cost management. In: FIG working week 2012 (conference), knowing to manage the territory, protect the environment, evaluate the cultural heritage, pp 1–7. Rome, Italy, 6–10 May 2012

Andrea E (2019) Effects of industry 4.0 on reshoring investments—Hungarian experiences. Centre for Economic and Regional Studies HAS. Institute of world economics working paper 251(2019). pp 1–51

Araszkiewicz K (2017) Digital technologies in facility management—the state of practice and research challenges. In: Creative construction conference 2017, 19–22 June 2017, Primosten, Croatia

Asuquo HO, Oladokun MG, Adelakun AD (2020) Digital transformation of the construction industry under the fourth industrial revolution. J Contemp Res Built Environ 4(1&2):95–113

Bahl M (2015) Asia rising: digital driving, cognizant centre for the future of work. Online available from http://www.futureofwork.com

Balfour Beatty (2017) Innovation 2050: a digital future for the infrastructure industry

Becerik B (2004) A review on past, present and future of web-based project management and collaboration tools and their adoption by the US AEC industry. Int J IT Archit Eng Constr 2/4

Berger R (2016) Digitalisation in the construction industry: a comprehensive guide to reinventing companies. www.rolandberger.com. Retrieved 3 Sept 2018

Bhuiyan AB, Ali MJ, Zulkifli N, Kumarasamy MM (2020) Industry 4.0: challenges, opportunities, and strategic solutions for Bangladesh. Int J Bus Manage Future 4(2):41–56

Bougourd J, Zuo J, Zillante G (2020) Factors influencing adoption of digital technologies in facility management. Facilities 38(5/6):285–302. https://doi.org/10.1108/f-02-2019-0028

Bröchner J, Haugen T, Lindkvist C (2019) Shaping tomorrow's facilities management. Facilities 37:366–380

Castagnino S, Rothballer C, Gerbert P (2016) What's the future of the construction industry? World Economic Forum. Retrieved from https://www.weforum.org/agenda/2016/04/building-in-thefourth-industrial-revolution/ on Feb 2019

CGI (2017) Industry 4.0 makes your business more competitive. Visit cgi.com or contact us at info@cgi.com for more information. © 2017 CGI Group Inc

Christensen CM, Raynor M (2003) The innovator's solution: creating and sustaining successful growth. Harvard Business School Press, Boston, Mass, pp 177–210

Crnjac, M, Veža, I, Banduka, N (2017). From concept to the introduction of Industry 4.0. Int J Ind Eng Manage 8(1):21–30. http://dx.doi.org/10.24867/IJIEM-2017-1-103

Dale J (2007) Innovation in construction: ideas are the future. CIOB Survey. www.Ciob.org.uk

Dall'omo S (2017) Driving African development through smarter technology. African Digitalisation Maturity Report. pp 1–45

Deloitte (2017) The fourth revolution is now: are you ready? Future of Operations. www.deloitte.com/futureofoperations. © 2017Deloitte LLP. All rights reserved

Deloitte (2020) 2020 commercial real estate outlook. https://www2.deloitte.com/us/en/pages/real-estate/articles/commercial-real-estate-outlook.html

Deloitte (2021) Real estate predictions 2021: adapting to the next normal. https://www2.deloitte.com/content/dam/Deloitte/us/Documents/real-estate/us-real-estate-predictions-2021.pdf
Egilmez G, Lützkendorf T, Reichardt A (2020) Digitalization in facility management: a case study of smart building operations. J Inf Technol Constr 25:321–337. https://doi.org/10.36680/j.itcon.2020.021
Eisenhardt KM (1989) Building theories from case study research. Acad Manage Rev 14(4):532–550
EMC (2016) Competing for digital customers: why companies must embrace digital transformation now. EMC Corporation. www.emc.com
Ernest and Young (2015) Risk and opportunity in an increasingly digital world, insurance governance leadership network. Tapestry Networks Inc, 2015. www.tapestrynetworks.com. Retrieved 13 Sept 2018
European Commission (2018) European commission digital strategy—a digitally transformed, user-focused and data-driven Commission. European Commission Publications
Eze NM, Ayigbe SF, Eberechi OO, Jordan N (2015) Information and communication technology applications in small and medium scale enterprises (SMEs) funding in Nigeria: an impact assessment. Int J Finance Acc 4(5):293–303
Ezeokoli FO, Okolie KC, Okoye PU (2016) Digital transformation in the Nigeria construction industry: the professionals' view. World J Comput Appl Technol 4(3):23–30
Fadahunsi JO, Utom JA, Ochim MR, Ayedun CA, Oloke OC (2019) Benefits of the adoption of facilities management practices in tertiary institutions: a case study of covenant university. In: IOP conference series: materials science and engineering, vol 640
Fitzgerald M, Kruschwitz N, Bonnet D, Welch M (2013) Embracing digital technology. Research report, MIT Center for Digital Business, MIT Sloan School of Management, Capgemini Consulting. MIT Sloan Manage Rev 55(2):1
Fuchs M, Müller C, Schäfer M (2020) Towards a comprehensive framework for digitalization in facility management. Procedia CIRP 97:127–132. https://doi.org/10.1016/j.procir.2020.02.023
García de Soto B, Agustí-Juan I, Joss S, Hunhevicz J (2022) Implications of construction 4.0 to the workforce and organizational structures. Int J Constr Manage 22(2):205–217. https://doi.org/10.1080/15623599.2019.1616414
Gerbert P, Castagnino S, Rothballer C, Renz A, Filitz R (2016) Digital engineering and construction: the transformative power of building information modelling. The Bolton Consulting Group, 2016 Report
Guest G (2012) Applied thematic and analysis. Sage, Thousand Oaks, California, p 11
Halim HC (2010) Assessment of the application of information and communication technology in real estate practice (a case study of Lagos Metropolis). MSc. Thesis, Department of Estate Management, University of Nigeria, Enugu
Helfert M, Borras J, Schröder C (2020) The impact of digitalization on facility management: an exploratory study. Int J Inf Syst Project Manage 8(3):5–22. https://doi.org/10.12821/ijispm080301
Himang C, Ocampo L, Obiso JJ, Bongo M, Caballes SA, Abellana DP, Deocaris C, Ancheta R (2020) Defining stages of Industry 4.0 adoption via indicator sets. Eng Manage Prod Serv 12(2):32–55
Ibem EO, Laryea S (2014). Survey of digital technologies in procurement of construction projects. Autom Constr 46:11–21
Ibrahim S, Birshir I (2012) Review of using building information modeling in Nigerian construction industry. J Environ Sci Policy Eval 2(2):221–230
Jadhav VV, Mahadeolar R, Bhoite SD (2019) The fourth industrial revolution (I4.0) in India: challenges and opportunities. Published in Int J Trend Sci Res Dev. ISSN: 2456-6470. Special issuelfostering innovation, integration and inclusion through interdisciplinary practices in management. pp 105–109
Kagermann H, Walister W, Helbig J (2013) Recommendations for implementing the strategic initiative Industrie 4.0: final report of the Industrie 4.0 working group

Koch C, Hansen GK, Jacobsen K (2019) Missed opportunities: two case studies of digitalization of FM in hospitals. Facilities 37:381–394

Kumar R (2011) Research methodology is astep-by-step guide for beginners, 3rd edn. SAGE Publications. ISBN978-1-84920-300-5. ISBN 978-1-84920-301-2(pbk)

Kundishora SM (2006) Partnership framework for ICT infrastructure development in Africa at world congress on ICT for knowledge society, Seoul

Lee JY, Irisboev IO, Ryu Y-S (2021) Literature review on digitalization in facilities management and facilities management performance measurement: contribution of industry 4.0 in the global era. Sustainability 13:13432. https://doi.org/10.3390/su132313432

Lekan A, Aigbavboa C, Babatunde O, Olabosipo F, Adediran C (2020) Disruptive technological innovations in construction field and fourth industrial revolution intervention in the achievement of the sustainable development goal 9. Int J Constr Manag. https://doi.org/10.1080/15623599.2020.1819522

Li X, Li N, Li H, Wang X (2021) Factors affecting the digital transformation of facility management: evidence from China. J Clean Prod 310:127399. https://doi.org/10.1016/j.jclepro.202

Lin YC, Hsu YT, Hu HT (2022) BIM model management for BIM-based facility management in buildings. Adv Civ Eng 2022:1–13

Ling YM, Hamid NAA, Chuan LT (2020) Is Malaysia ready for industry 4.0? Issues and challenges in manufacturing industry. Int J Integr Eng 12(7):134–150

Mäkinen T (2017) Strategising for digital transformation: a case study of digital transformation process in the construction industry. Department: Industrial Engineering and Management. School of Science, Aalto University

Mansour H, Aminudin E, Tahseen M (2021) Implementing industry 4.0 in the construction industry-strategic readiness perspective. Int J Constr Manage https://doi.org/10.1080/15623599.2021.1975351

Miehlbradt L, Schmittner C, Krcmar H (2021) Digitalization of facility management: a systematic literature review. Sustainability 13(2):667. https://doi.org/10.3390/su13020667

Mirarchi C, Pavan A, Marco FD, Wang X, Song Y (2018) Supporting facility management processes through end-users' integration and coordinated BIM-GIS technologies. Int J Geoinf 7:191

Moshood TD, Adeleke AQ, Nawanir G, Ajibike WA, Shittu RA (2020) Emerging challenges and sustainability of industry 4.0 era in the Malaysian construction industry. Int J Recent Technol Eng 9(1):1627–1634

Mukwawaya GF, Emwanu B, Mdakane S (2018) Assessing the readiness of South Africa for industry 4.0—analysis of government policy, skills and education. In: Proceedings of the international conference on industrial engineering and operations management Pretoria/Johannesburg, South Africa, Oct 29–Nov 1 2018

Mutesi ET, Kyakula M (2009) Application of ICT in the construction industry in Kampala. In: Advances in engineering and technology: contribution of scientific research in development: proceedings of the second international conference on advances in engineering and technology, Kampala. Makrere University, Tanzania

Neumann WP, Winkelhaus S, Grosse EH, Glock CH (2021) Industry 4.0 and the human factor—a systems framework and analysis methodology for successful development. Int J Prod Econ 233:1–16

OECD (2019) Going digital—making the transformation work for growth and well-being. OECD Publications. https://www-oecd-org.focus.lib.kth.se/going-digital/. Accessed 24 Sept 2019

Ofoefule C (2018) Digital transformation benefits, strategy and potentials for Nigeria and Africa. Hemingway Global Services Ltd—©2018 All Rights Reserved. www.hemingway.com.ng

Oke AE, Aghimien DO, Aigbavboa CO, Koloko N (2018) Challenges of digital collaboration in the South African construction industry. In: Proceedings of the international conference on industrial engineering and operations management Bandung, Indonesia, Mar 6–8 2018

Oke AE, Arowoiya AV, Akomolafe OT (2020) Influence of the Internet of Things' application on construction project performance. Int J Constr Manag. https://doi.org/10.1080/15623599.2020.1807731

Oladokun, MG, Asuquo, HO, Adelakun, AD (2021) Benefits of digital transformation of the construction industry under the fourth industrial revolution. J Environ Des 16(1):44–54
Oladapo AA (2007) An investigation into the use of ICT in the Nigerian construction industry. Electron J Inf Technol Constr 12:261–277
Olaitan OO, Issah M, Wayi N (2021) A framework to test South Africa's readiness for the fourth industrial revolution. S Afr J Inf Manage 23(1):1–10
Olufemi A (2020) National digital economy policy and strategy (NDEPS): 2020–2030. DTTWG%20Training_National%20Digital%20Economy%20Policy%20and%20Strategy.pdf
Osunsanmi TO, Aigavboa C, Oke A (2018) Construction 4.0: the future of the construction industry in South Africa. World Acad Sci Eng Technol Int J Civ Environ Eng 12(3):206–212
PWC (2016) Industry 4.0: building the digital enterprise. In: Global Industry 4.0 Survey what we mean by Industry 4.0/Survey key findings/Blueprint for digital success. www.pwc.com/industry40
RICS (2020) PropTech and the future of property. https://www.rics.org/globalassets/rics-website/media/knowledge/research/uk/rics-proptech-and-the-future-of-property.pdf
Rojko A (2017) Industry 4.0 concept: background and overview. https://doi.org/10.3991/ijim.vlli5.7072. Retrieved 10 Sept 2018
Rouse M (2017) Meaning of digitalisation. Retrieved from http://whatis.techtarget.com/definition/digitization on Feb 2019
Saunders M, Lewis P, Thornhill A (2009) Research methods for business students, 5th edn. Pearson, England, pp 1–617. Retrieved from http://ebooks.narotama.ac.id/files/Research Methods for Business Students (5th Edition)/cover %26 Table of contents-Research Methods for Business Students (5th Edition).pdf
Schwab K (2016) The fourth industrial revolution. World Economic Forum 91–93 route de l a CapiteCH-1223 Cologny/Geneva Switzerland. ISBN-13:978-1-944835-01-9. ISBN-10: 1944835016. REF: 231215. www.weforum.org
Strange R, Zucchella A (2017) Industry 4.0, global value chains and international business. Multinatl Bus Rev 25:174–184
The World Bank Group (2019) Digital development. The World Bank Group online resources. https://www.worldbank.org/en/topic/digitaldevelopment/overview#1. Accessed 24 Sept 2019
Vaidya S, Ambad P, Bhosle S (2018) Industry 4.0—a glimpse. Procedia Manuf. In: 2nd international conference on materials, manufacturing and design engineering (iCMMD2017), 11–12 Dec 2017, MIT Aurangabad, Maharashtra, INDIA 20. pp 233–238. https://doi.org/10.1016/j.promfg.2018.02.034
Verdantix (2020) Smart building technologies 2020: market size and forecast. https://www.verdantix.com/research/report/smart-building-technologies-2020-market-size-and-forecast
Wang T, Ali AS, Au-Yong CP (2022) Exploring a body of knowledge for promoting the building information model for facility management. Ain Shams Eng J 13(2022):1–14
Wijesinghe SMA (2011) From interpretivism to positivism: a methodological approach to the research in human resource management (HRM). In: 8th international conference on business management
Wix J, Nicholas N (2008) Study into the business case for interoperable building information modeling (BIM). Int J Constr Inf Technol
Wong JKW, Ge J, He SX (2018a) Digitalisation in facilities management: a literature review and future research directions. Autom Constr 92:312–326
World Economic Forum (2019) Annual report 2018–2019. World Economic Forum publications
World Economic Forum (2020) Africa needs digital skills across the economy—not just the tech sector. https://www.weforum.org/agenda/2020/10/africa-needs-digital-skills-across-the-economy-not-just-tech-sector/
Yin RK (2003) A review of case study research: design and methods. In: Applied social research methods, 3rd edn, vol 5. Sage Publications, London, pp 1–181

Chapter 4
Improving Facility Management Performance in Public Higher Education Institution Buildings in Nigeria

Andrew Ebekozien, Clinton Ohis Aigbavboa, Mohamad Shaharudin Samsurijan, and Mohd Isa Rohayati

Abstract Studies have shown that effective facility management (FM) is successful is key to sustainable building performance. It enhances the functionality and sustainability of public higher education institution buildings (PHEIBs). However, the performance of FM in Nigerian PHEIBs has yet to receive in-depth study. This study offers a collective insight into hindrances facing FM performance and suggests measures to improve facilities' performance within the perception of public higher institution buildings in Nigeria. Given the unexplored dimension of the issue, data were collected via face-to-face interviews across selected public higher institutions in Nigeria. Twenty-six participants were interviewed and collated data analysed via a thematic method. Findings show that facilities management performance in Nigerian PHEIBs is characterised by passive actions in the twenty-first century driven by artificial intelligence and less than a decade to the 2030 Agenda for Sustainable Development Goals (SDGs). As part of the study's implications, findings will strengthen and offer a robust insight into a proposed integrated framework for facilities performance in Nigerian PHEIBs, especially concerning functionality and sustainability. The outcome will activate transformation and sustainable development to benefit higher education institutions and humanity.

A. Ebekozien (✉) · C. O. Aigbavboa
Department of Construction Management and Quantity Surveying, University of Johannesburg, Johannesburg, South Africa
e-mail: ebekoandy45@yahoo.com

A. Ebekozien
Department of Quantity Surveying, Auchi Polytechnic, Auchi, Nigeria

A. Ebekozien · M. S. Samsurijan
Development Planning and Management, School of Social Sciences, Universiti Sains Malaysia, Pulau, Pinang, Malaysia

M. I. Rohayati
National Higher Education Research Institute Centre, Universiti Sains Malaysia, Pulau,, Pinang, Malaysia
e-mail: rohayati@usm.my

S. P. Akinbogun et al. (eds.), *Facility Management practices*,
https://doi.org/10.1007/978-3-031-65321-6_4

Keywords Development · Facilities · Higher education institution · Management · Performance

4.1 Introduction

Higher education institutions are essential to national capacity and link with global education trends. This knowledge can enhance economic growth and accelerate sustainable development. It is achievable in an enabling environment supported by effective and operative facilities management (Odediran et al. 2015). A recent trend in the built environment has embraced facilities management as one of the youngest disciplines within the built environment industry. It is regarded as a newcomer in architecture, engineering, real estate, quantity surveying, and construction. The publication of ISO 41001 *Facilities Management—Management Systems* in 2018 has enhanced the standard internationally and places it on par with the first-rate number of high-profile industries (Atkin and Brooks 2021). Practitioners and scholars have viewed the term "Facilities Management" from different perceptions but tailored toward the same aims. The International Facility Management Association IFMA (2006) described FM as intergrading, coordinating, and managing interconnected "people, process, and place" matters and their purposes within a company or an association. Amaratunga and Baldry (2000) believe facilities management is accountable for organising all matters connected to planning, designing, and managing structures. This includes their systems, equipment, and fittings to improve the firm's capacity to compete in a changing environment. Also, the term is ascribed as a method that ensures that building and other technical mechanisms support the processes of a firm and integrate an approach to operating, maintaining, improving, and adapting the building and infrastructure of a firm. It will develop an environment that supports the firm's main aims (Atkin and Brooks 2000, 2021; Fenker 2004). Asiabaka (2008) described FM in connection with educational facilities as the tender of scientific approaches in administering the physical environment of academic learning for achieving educational goals.

Studies (Price 2003; Marmolejo 2007; Odediran et al. 2015) identified the learning environment as influencing effective teaching and learning. It implies that appalling facilities impede the achievement of this goal (effective teaching and learning). This is complicated if the facilities are ageing as identified by Marmolejo (2007) in some developed countries and influence academic and research performance. Regardless of the institution and building, higher education institutions may be experiencing change related to budget cuts and financial crises to manage these facilities. In developing countries such as Nigeria, to mitigate this crisis associated with the current ageing/fading inadequate infrastructure in higher education institutions and government budget deficit, though far below the UNESCO recommendation (Enefola 2016) demand the need for alternate infrastructure financing paths and initiatives (Gbadegesin and Aluko 2014). Still, infrastructure is getting more deplorable (National Universities Commission (NUC) 2006; Odediran et al. 2015; Amoo 2018; Olufemi

2020). Thus, Odediran et al. (2015) emphasised the need for an effective higher education institution to retain effective teaching and learning outputs. Effective facilities management has been proven successful in sustainable building performance. This technique can enhance the functionality and sustainability of public higher education institution buildings (PHEIBs). In Nigeria's context, a few studies (Ikediashi et al. 2012; Gbadegesin and Aluko 2014; Odediran et al. 2015; Abdullahi and Yusoff 2019) have been conducted on facilities management, not related to facilities management performance in public higher education institution buildings with the exemption of Abdullahi and Yusoff (2019). Abdullahi and Yusoff (2019) focused on how facilities performance influences student satisfaction in Northern Nigerian universities. Thus, how far regarding facilities management performance on Nigerian PHEIBs is yet to receive in-depth studies. This chapter investigates collective insight into hindrances facing FM performance and proffers measures to improve facilities' performance within the perception of Nigerian public higher institution buildings. The study will achieve the aim of the chapter via the following stated objectives:

(i) Investigate the perceived hindrances facing FM performance in Nigerian PHEIBs.
(ii) Suggest feasible measures to improve facilities management performance within the perception of public higher institution buildings in Nigeria.

This chapter focused on the main research question—What are the encumbrances facing FM performance, and how to improve efficiency regarding public higher education institution buildings in Nigeria? The chapter intends to investigate collective insight into hindrances facing FM performance. Also, suggest measures to improve facilities' performance within the perception of Nigerian public higher institution buildings. This chapter is divided into seven sections. The first section focuses on the introduction, including the chapter's main objectives. The second section highlights previous related literature. This includes facilities management performance, infrastructure development in higher institutions, and hindrances facing facilities management performance in Nigeria. Next is collecting data from 26 knowledgeable participants via in-depth virtual interviews nationwide. The fourth section is the analysed findings via thematic and discussion with past related results. The fifth section focuses on the study's contribution to theory and practice. The limitations and areas for further study follow this. Lastly, the concluding section comprises the chapter's conclusion and recommendations.

4.2 Facilities Management Performance

Facilities management's key function is coordinating support services, the firm's requests, main activities, and personnel (Chitopanich 2004). It is noted that structured and planned FM may advance the physical presentation and presence of a building and its system and increase the operators' satisfaction level. Also, to advance the competence with which the structure is sustained and managed (Lavy 2008). The desire for better facilities management performance has become pertinent because buildings and equipment components are erudite (Amaratunga and Baldry 2002; Odediran et al. 2015). Yoon and Cha (2017) found quick responsiveness to an emergency, service reliability, tenants' safety, work execution, and customer satisfaction critical to successful FM performance. This study cannot overstate the specific purpose of facilities management to establish their performance to the students and staff in an academic environment. It is acknowledged that an interrelation within the firm context offers the catalyst for an enhanced performance that enables teaching and learning and increases instructional efficiency for students and staff. It is because the UK-based Commission for Architectural and Built Environment (2005) relates the competent and real facilities management of educational institutions to the overall service delivery in higher education institutions. Belcher (1997) opined that higher education institutions' facilities management performance services had become a concern to concerned authorities. The aim is to enhance the effectiveness of the educational provision and maximise value for money, especially for low-income countries with a poor budget for the education sector. This submission corroborated Preiser (1995) and affirmed that higher education institutions are trusted to utilise public funds prudently. The rising operational costs and increasing user expectations make many higher education institutions integrate FM into their managerial mission. The outcome has the massive expansion of knowledge transfer and economic use of their facilities (Clarke 1997).

In Nigeria, facilities management has wide applications in public agencies, companies, and non-profit organisations, including higher education institutions, which have understood that managing these elements within the conventional or traditional systems is disappointing (Sani 1998). Adewunmi et al. (2009) asserted that building management practice to secure optimum returns has been prevalent and validated that the practice, generally, has been focused on building management. In higher education institutions, FM is enabled by the management section and, in some cases, merged with the work's directorate or physical planning section (Ikediashi et al. 2012). Poor infrastructure facilities are not peculiar to Nigeria. Moore and Finch (2004) identified developmental problems like globalisation, ICT, staff prospects, provincial economy, property market, and procurement methods as issues influencing facilities management growth in Southeast Asia. Inadequate funding, poor awareness, lax regulation, etc. (Ikediashi et al. 2012), and overcrowding (Adegbite 2007) were major hindrances to FM in Nigerian higher education institutions. These factors may hinder FM performance in Nigerian PHEIBs. Other factors include facilities that do not receive explicit management attention, as revealed in the research conducted by Graham Bannock and Partners and Workplace Management, cited in Then (1999). Rose et al. (2007) identified time wastage in initial planning, design, and construction costs of a facility and frequent with the United States higher education institution-building projects. Also, Asiabaka (2008) found that apart from higher education institution-building ageing, they need to progress with digital innovation and societal changes. Nigerian higher education institution buildings are not exempted. Asiabaka (2008) discovered that managers or relevant officials overlook maintenance plans. An estimated US$66 billion is required yearly for infrastructure development in African nations, including Nigeria (Wentworth and Makokera 2015). Embracing facilities management performance on higher education institution buildings may reduce funding because fund accessibility from international partners, alliances with development finance institutions, and private investors are getting more stringent. Nigeria's case is tougher because of alleged disregard for some past concession construction projects (Elebiju and Ilesanmi 2020).

In line with Amaratunga and Baldry (2000), the balanced scorecard (BSC) framework was modified in Nigeria's context, as presented in Fig. 4.1. The tool offers managers and policymakers the instrumentation to navigate future competitive success via four critical perspectives. They include internal business process, customer, financial, and learning and growth viewpoints, as presented in Fig. 4.1. Referring to Fig. 4.1, the three components of business stakeholders are captured—shareholders, customers, and employees. One attribute of the BSC approach to improving FM performance is that it reflects the facilities management firm's attributes, goals, and critical success factors. It is a new approach to performance evaluation in facilities management and possibly the first in Nigeria's context regarding higher education institution building's facilities management performance. The approach considers that no single performance indicator can capture the full complexity of a firm's performance (Amaratunga and Baldry 2000). One key to the BSC method is firm causal relationships, as shown in Fig. 4.1. The essence of the proposed framework is to devise a performance evaluation framework for facilities management. This study focused on public higher education institution buildings. The framework will assist in appreciating the benefits of facilities management performance of higher education institution buildings, such as communication and teamwork, commitment, and feedback and learning. Also, the framework intends to integrate facilities performance in Nigerian PHEIBs, especially concerning functionality and sustainability, and change the narrative from passive to active actions in the twenty-first century driven by artificial intelligence (digital technology). Achieving SDGs associated with quality education in 2030 may become a reality.

4.2.1 Infrastructure Development in Higher Institutions

Globally, higher education is a critical instrument for national expansion and growth. It is a platform for creating higher-level skills and proficiencies tailored toward knowledge economies (Abdullahi and Yusoff 2019). Johnston (2007) posited that higher education institutions are agents of change, integrated national development, and mechanisms for actualising collective aspirations. Higher education institutions can achieve these aspirations via teaching, curriculum enhancement, and educational research. Nigeria is not exempted from this actualisation, calling for concerns regarding feasible policies and programmes to provide the crucial mass skills and education the populace requires for genuine endogenous sustainable development. In Nigeria, public higher education institutions comprise post-secondary institutions

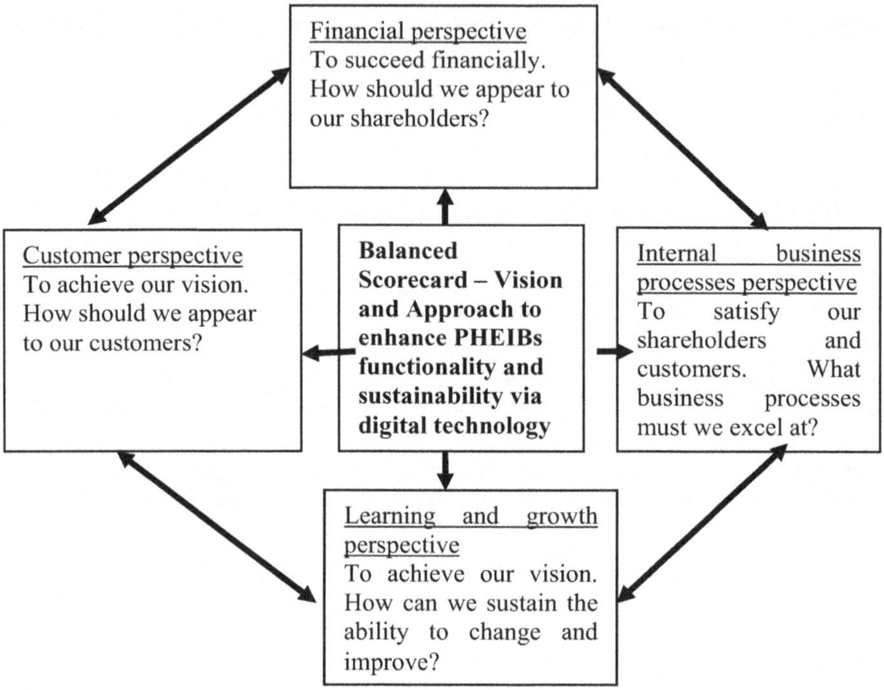

Fig. 4.1 Proposed modified Nigeria's BSC framework—four perspectives. *Sources* Kaplan and Norton (1996) and Amaratunga and Baldry (2000, p. 297)

or tertiary institutions owned by the governments (federal or state) as a document by the National Policy on Education (NPE as cited in Abdullahi and Yusoff 2019). The student enrolment standard in higher education institutions per 100,000 residents shows an over sixfold increase, from 13 million in 1996 to 102 million in 2003 (UNESCO 2006). The gap is wider between developed and developing countries. It shows that the accessibility and availability of resources for higher education and research are limited in developing countries. They may have contributed to an increase in socio-economic stratification. The governments are accountable for ensuring access to basic infrastructure facilities, including public higher education buildings. Many developing countries' governments need help tackling the infrastructure backlogs (Wentworth and Makokera 2015). The critical role of infrastructure in transforming higher education institutions for sustainability and national growth cannot be over-emphasised (Enefola 2016).

Also, public higher education institutions need to contribute to economic development. Physical infrastructure availability and accessibility are critical in achieving this goal. Higher education institutions with adequate physical infrastructure could help develop collective innovation assets (Anderton 2016). In many developing countries, meeting the minimum infrastructure development in higher education institutions are critical problems that have lingered for a long time. Atuahene (2014) found inadequate funding as one critical problem and a threat to Ghanaian higher education development. In Nigeria, most higher education institutions are working with low-quality facilities that cannot accommodate student enrolments yearly demands (Abdullahi and Yusoff 2019). The Needs Assessment Report of Nigerian public universities (2012) acknowledged that there are deteriorating and insufficient physical facilities, worsened by lecture rooms/theatres congestion and shortage of accommodation facilities. This is a source of concern as the functionality and quality of academic facilities may impact graduates' performance in practice. In response to the Needs Report, the government established the Tertiary Education Trust Fund (TETFund). The TETFund Act of 2011 repealed ETF (TETFund 2011). As part of the agency's responsibilities, funds are disbursed for infrastructure, human capacity development, institutional equipment, etc. (Anaehobi and Agim 2019). Yet, records show lacuna in existing aged and new infrastructure regarding maintenance. Thus, embracing effective facility management (FM) to enhance and sustain higher education institution buildings' performance cannot be overstated. The technique may enhance the functionality and sustainability of public higher education institution buildings. Gbadegesin and Aluko (2014) and Enefola (2016) discovered a lack of infrastructure in Nigeria's educational sector because of inadequate funding. Enefola (2016) found that from 1999 to 2014, the budgetary allocations were far below the recommended 26% benchmark by UNESCO. Amoo (2018) reported that in 2019, the budgetary allocation was 7.02 and 5.6% for 2021 (Olufemi 2020).

4.3 Research Method

The researchers adopted a qualitative research approach for this research. The research aligns with the interpretivism epistemological idea. This approach makes the study unexplored in Nigeria's context. This aligns with Amaratunga and Baldry (2000). They conducted a similar study using interviews via a semi-structured questions approach. Concerning the ontological point of view, the study aligns with constructivism. This implies that social entities and their meaning are not independent of social actors (Bryman 2016). The study collected data via semi-structured virtual interviews across the six geo-political zones. A semi-structured interview was adopted because it allows for an investigatory study (Saunders et al. 2012). Hence, it allows the investigators to probe the participants further. The 26 participants were directors, top senior officers in the physical planning/maintenance/estate and work department of the selected higher education institutions (universities and polytechnics), management staff of selected construction companies, maintenance/facility consultants, and policymakers from government agencies, as presented with detailed background in Table 4.1. The study achieved this via a purposive sampling technique, a non-probability form of sampling in which participants are sampled purposefully to answer the research questions (Bryman 2016; Ebekozien et al. 2022; Ibrahim et al. 2022). Purposeful sampling was utilised to get information from specific target groups favourably placed based on their position (Sekaran and Bougie 2016; Jaafar et al. 2021). Participants were from 12 (five polytechnics and seven universities) public higher education institutions. The study ensured that at least two higher educational institutions were covered in each Nigerian geo-political zone for a good representation and achieved saturation. The interviews took an average of 40 min and were conducted from August to November 2021. The study concealed the personal details of the participants. This is part of the study's ethical rule. Table 4.1 shows that those engaged are knowledgeable in higher education institutions FM in Nigeria with at least 15 years of work experience in their respective field. Appendix presents the cover letter and sample of the semi-structured interview questions.

In generating the codes, this study used thematic analysis and manually analysed them. The investigators read twenty-six documents multiple times and played the role of the coders to describe the participants' viewpoints regarding the phenomenon. It aligns with Omotayo et al. (2020), that employed the same method to produce the initial coding method for their study. Two coding phases were utilised in line with Ebekozien (2019). The first was open coding (Saldana 2015). The last phase of this study involves utilising the sub-themes that emerged from the first phase to re-read the transcript precisely and identify the new constructs. The researchers created two themes from the sub-themes with the help of the thematic approach. Member checking, triangulation, and researcher reflexivity were adopted as the validity approach to the collected data (Creswell and Creswell 2018; Ebekozien 2020a, b). The research adopted themeing, invivo, and narrative techniques in the data coding (Saldana 2015). Sixty-nine codes were generated and re-grouped into six categories. Two themes emerged from the six categories.

4.4 Findings and Discussion

In Nigeria, the facilities management performance of public higher education institution buildings has become germane. This is because the role of functional and sustainable infrastructure in the economic development of countries, especially in higher education institutions, cannot be overstated. There is a paucity of literature regarding this matter in developing countries such as Nigeria. It has become pertinent to investigate the perceived hindrances facing FM performance and proffer measures to improve facilities' performance within the perception of public higher institution buildings in Nigeria. The main aim was achieved via two themes as presented. Figure 4.2 presents the thematic network of the themes and the connection as part of the study's implications.

Theme One: Perceived hindrances facing FM performance

Researchers agree that FM is critical to manufacturing and service firms' effectiveness and competitiveness. However, issues may hinder FM performance in higher education institution buildings and still need to be developed. This theme allows the identification of perceived hindrances facing FM performance in Nigeria's higher education institution buildings. Ten perceived hindrances from the analysed data, as

Table 4.1 Summary of interviewees' description

ID	Organisation	Location	Years of experience	Rank/Firm
P1	Federal University	North Central	25 years	Deputy Director, Physical Planning Dept
P2	State Polytechnic		21 years	Senior staff in Maintenance Dept
P3	Medium construction company		28 years	Head, Operation
P4	State University	North East	20 years	Senior Staff, Physical Planning Dept
P5	Federal Polytechnic		21 years	Senior Staff, Physical Planning Dept
P6	Medium construction firm		22 years	Operational Head
P7	Federal University	North West	24 years	Deputy Director, Physical Planning Dept
P8	State University		28 years	Director, Physical Planning Dept
P9	Medium construction firm		23 years	Director
P10	State University	South East	20 years	Senior staff, Physical Planning Dept
P11	Federal Polytechnic		23 years	Senior staff, Physical Planning Dept
P12	Medium construction firm		35 years	Director
P13	Federal University	South West	28 years	Deputy Director, Physical Planning Dept/Academic Staff (Asso. Prof.)
P14	State Polytechnic		27 years	Director, Physical Planning Dept
P15	Large construction firm		32 years	Manager, Engineering Department
P16	State University	South-South	29 years	Director, Physical Planning Dept
P17	Federal Polytechnic		21 years	Senior staff, Physical Planning Dept
P18	Large construction firm		20 years	Head, Maintenance Contract
P19	FM consultant	Lagos	16 years	Ass. Director
P20	FM consultant		22 years	Operation Manager
P21	FM consultant	Abuja	17 years	Principal Partner
P22	FM consultant		19 years	Managing Partner

(continued)

Table 4.1 (continued)

ID	Organisation	Location	Years of experience	Rank/Firm
P23	Tetfund staff	Abuja	15 years	Senior staff
P24	Tetfund staff	Abuja	18 years	Senior staff
P25	NUC staff	Abuja/Kaduna	20 years	Senior staff, a government agency under the Federal Ministry of Education
P26	NBTE staff		22 years	Senior staff, a government agency under the Federal Ministry of Education

presented in Fig. 4.2, summarise the issues. This includes using traditional systems, facilities that do not receive explicit management attention, alleged corruption in the system, time wastage on planning, design, and construction costs of a facility, and ageing building not revitalised. Others are the absence of a maintenance plan for old and newly constructed buildings, obsolete educational buildings, inadequate physical facilities, low maintenance budget, and fund accessibility issues from international partners. Of the ten hindrances, the traditional system, physical facilities not receiving management attention, alleged corruption, and absence of maintenance plan for old and new buildings cut across the participants. P1 says, "…. *the hindrances are everywhere. Apart from the inadequate physical facilities, I may take you to the office I share with a colleague who is a senior staff like me. Most of the buildings are 'architecturally obsolete.' There is nothing to contribute to functional learning except if upgraded….*" The findings agree with Asiabaka (2008) and Odediran et al. (2015). They affirmed that many Nigerian higher education institution buildings could not contribute to functional learning because they are inadequate, obsolete, and need urgent attention. Findings agree that the issue of corruption across Nigeria's system is hampering policies and programmes. Participant P19 says, "…. *No maintenance is conducted on these buildings, though we hear of the yearly budget for higher education institution-building maintenance, 'year-in-year-out.' There is no proper channel to report the case. It is an allegation, but I believe with proper investigation, we can improve the system across the sectors….*"

Participant P13 says, "…. *Most times I visit other institutions for national engagement (accreditation), I wonder why managers (VC and Rectors) of these higher institutions, with the exemption of an extreme few, show less concern about maintenance culture. I do not want to believe they don't know that infrastructure development in public higher education institutions can influence the externalities and attract global research grants and outputs……*" Findings agree with Enefola (2016) and Benneworth and Fitjar (2019). Enefola (2016) affirmed that infrastructure in higher education institutions could enhance sustainability and promote national growth and development for the school and the nation. Benneworth and Fitjar (2019) corroborated Enefola's assertation that higher institutions contribute to worker market

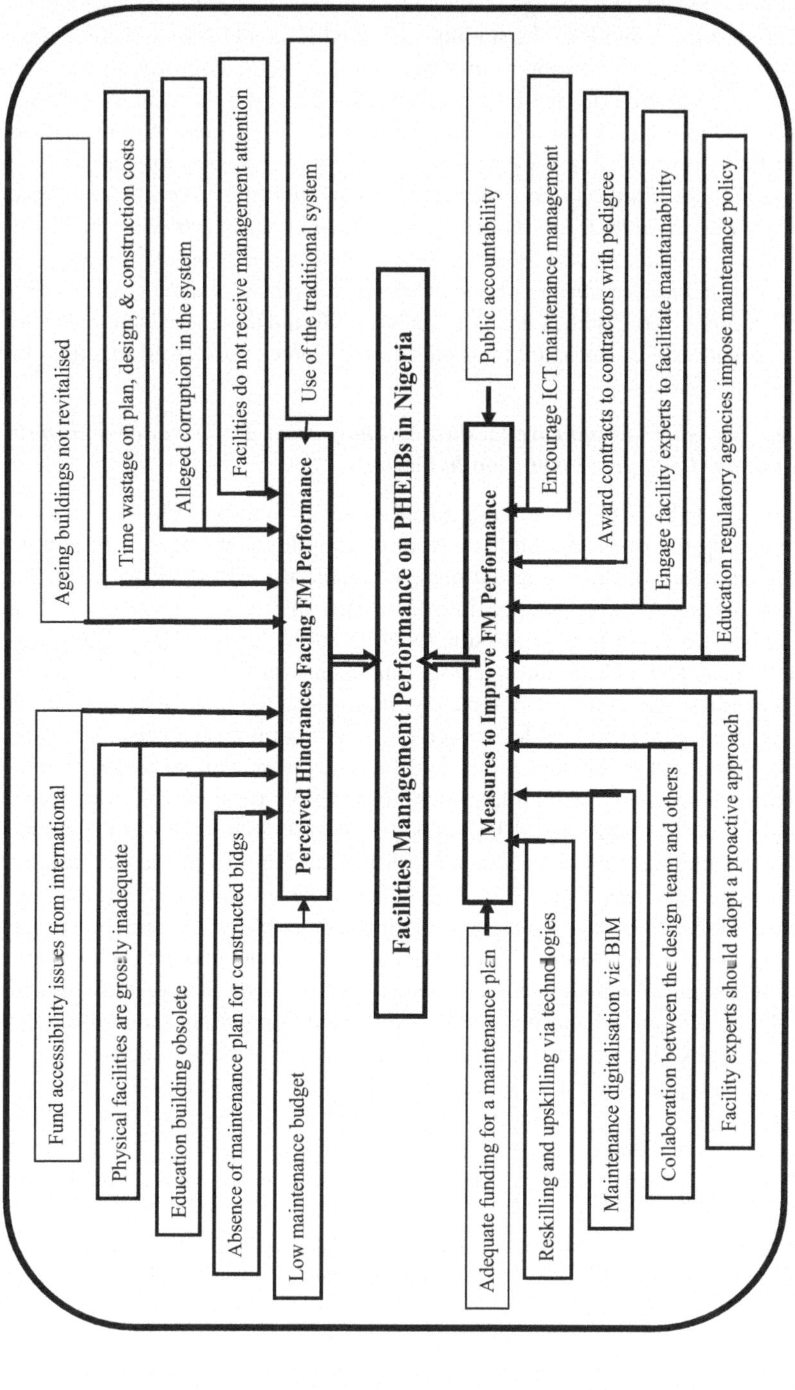

Fig. 4.2 Thematic network of facilities management performance on public higher education institution buildings in Nigeria

upskilling. "*.... our team needs to change their ways of thinking and doing things. We are in the digital world, and change is dynamic, but here, the 'old gods' want things done in their ways....*' said P4. The findings agree with Sani (1998) and Adewunmi et al. (2009) regarding the disappointing traditional system. They found the wide application of FM in Nigeria, but the approach has been dominant and obsolete. It implies that the focus has been on building management to secure optimum returns. Participant P20 says, "*.... it's high time our policymakers and managers of these higher education institutions knew that educational facilities, like other material resources, are consumable. It is either there is a plan in place to replace or revitalise after a certain maximum period. If not, living in them becomes more dangerous, as seen in many higher institutions. I don't want to mention names and the specific buildings.......*" Findings agree with Asiabaka (2008) and found that there is no provision for the maintenance of constructed buildings and, in some cases, overlooked by the institutions' managers.

Theme Two: Measures to improve facilities management performance within the perception of public higher institution buildings

Theme two offers the platform to suggest measures to improve FM performance within the perception of public higher institution buildings in Nigeria. Figure 4.2 presents the summary of the suggested measures to improve FM performance. The ten emerged measures were grouped into five. This includes (i) statutory obligations—National Universities Commission (NUC) and the National Board for Technical Examination (NBTE) should make it compulsory for the institution to have a maintenance policy as a requirement for accreditation, and facilities managers should adopt a proactive approach; (ii) design stage—a collaboration between the design team, contractors, and maintenance unit and facilities management experts should be engaged to facilitate maintainability after construction; (iii) construction stage—digitalisation such as BIM should be encouraged, and contractors with pedigree engaged; (iv) managing maintenance unit—reskilling and upskilling via maintenance technologies and innovation and ICT maintenance management encouraged; and lastly, (v) maintenance budget—adequate funding and public accountability (financial management and transparency). Findings agree that higher education institution buildings (old or new) should be revitalised and digitalised to be artificial intelligence (AI) friendly. Teaching and learning in the twenty first-century are tailored towards digitalisation. Thus, the government should digitalise the learning environment for efficiency and integrated service delivery (P1, P11, P17, P25, and P26). Findings agree with Araszkiewicz (2017) and found that digital techniques can promote smart facility management. Smart facility management in higher education buildings will enhance teaching, learning, and research. The outcome will improve performance and productivity.

Participant P19 says, "*.... regarding upskilling and reskilling of facilities management staff, many higher education institution managers (VCs and Rectors) are not willing to set a budget aside for upskilling and reskilling of maintenance section personnel regarding public FM, depriving these staff of knowing the current global minimum standard expectation in line with the best practices...... This is a critical* *issue, and our thinking should change to mitigate this issue....*" Findings agree with Ogungbile and Oke (2015) and recommend integrating facility managers into the design team to enhance and facilitate preventive and planned approachable maintenance mechanisms. FM practices have moved beyond just maintenance, and this narrative cannot change if the staffers are not subjected to global training to acquire the technique and knowledge (P12, P18, P21, P22, and P26). Participant P22 says, "*.... management team of these institutions should express the willingness to revamp these obsolete buildings in phases to meet with the 21st-century higher education institution buildings for teaching and learning* via *digitalisation.... How many campuses in Nigeria have internet (wifi) that students and staffers can access for free? How many buildings are powered* via *solar, promoting green buildings? These are some germane issues that should be addressed to enhance facilities management performance....*" Higher education institution buildings' sustainability and functionality are relevant in this twenty-first century for better student and staff productivity (P3, P6, P14, and P20). Higher education institution buildings should be designed and constructed strategically to proffer solutions to issues (enhance teaching and learning) and promote sustainability and functionality (P5, P18, and P23). This is pertinent, and digitalisation is key to driving this goal. Findings agree with Aboginije and Thwala (2020) and affirm that developing countries' fourth industrial revolution mechanisms could lead to higher infrastructure satisfaction.

Regarding National Universities Commission and the National Board for Technical Examination making it compulsory for institutions to have a maintenance policy as a requirement for accreditation, findings reveal that the facilities management institutional framework is germane and supported by the balanced scorecard approach adopted by Amaratunga and Baldry (2000). If well developed, it is an all-inclusive framework intending to integrate facilities performance in Nigerian PHEIBs, especially concerning functionality and sustainability, and change the narrative from passive to active actions in the twenty-first century driven by artificial intelligence (digital technology). The outcome can enhance achieving SDGs associated with quality education for all in 2030 and may become a reality. Participant P22 opines that BSC is a new mechanism for assessing facilities management. It has an organisational causal relationship, as shown in Fig. 4.1. Also, P22 and P23 emphasise exploring the benefits of digitalisation in facilities management. Findings agree with Wong et al. (2018) and found that digital facilities management will improve the accuracy of point cloud data for developing as-built models for existing facilities and generating geographic information system asset database integration. P23 says, "*.... many technologies such as 3D printing and BIM, components of construction 4.0 technology can be integrated into FM because of its benefits. This includes schedule performance, material efficiency, increased worker productivity, environmentally friendly maintenance design, and reduced human errors.......*" This will enhance cleaner activities and sustainable construction (P26).

4.5 The Study's Implications for Theory and Practice

From the reviewed literature, sustainable and functional higher education institution buildings can influence the teaching, learning, and research outputs. Studies have been conducted on facilities management and performance measurement, but none regarding possible hindrances facing facilities management performance on public higher education institution buildings in Nigeria. Thus, this study investigated collective insight into hindrances facing FM performance and suggested measures to improve facilities' performance within the perception of public higher institution buildings in Nigeria through the proposed modified framework. This research established a theoretical gap stemming from past similar policies concerning higher education institutions' facilities management performance to improve educational infrastructure development. Findings showed that effective facilities management could sustain building performance. Still, these facilities are facing challenges that need stakeholders' attention to enhance the functionality and sustainability of public higher education institution buildings. The study proposed a framework to enhance this goal. Also, the emerged thematic network of the major findings, as demonstrated in Fig. 4.2, is a component of the implications. Theoretically, this research investigated issues hindering facilities management performance on public higher education institution buildings in Nigeria and proffer measures to improve facilities management performance because of its benefits.

Concerning the study's practical implication, the study confirms that Nigeria's public higher education institution buildings need institutional synergy to explore possible measures to revitalise the obsolete buildings and make them sustainable and functional to meet the minimum global standards of educational facilities environment. The study's findings would assist and stir public higher education institutions, policymakers, and other stakeholders in connection with physical infrastructure development in public higher education institutions. Managers of public higher education institutions and policymakers should take advantage of the measures to improve FM performance on physical buildings. This study, by extension, will facilitate managers of higher education institutions to establish a facilities management performance framework to monitor and ensure that buildings are sustainable and functional within the context of the twenty first-century education environment. This forms the practical implication and developing countries' public higher education institutions with similar FM performance issues may adopt some of the study's recommendations.

4.6 Limitations and Future Areas of Study

The study recorded some limitations. First, this research employed a qualitative design that covered selected polytechnics and universities across six geo-political zones in Nigeria. Second, the restricted sample size used for the analysis is specific to the geo-political zones of Nigeria. However, based on the extensive literature review and the findings, the study shows that this does not negatively affect the results and may be generalised and adopted in other developing countries' public higher education institutions facing similar facilities management performance encumbrances. Also, though the study utilised a controlled sample size of 26 interviewees, the results were analysed with the extensive, robust reviewed literature, and saturation was accomplished. Thus, further studies are essential to validate the proposed framework with the emerged variables in the thematic network.

4.7 Conclusion and Recommendations

The chapter showed that effective facilities management is a successful tool to sustain building performance, but there are challenges. The study offers a collective insight into hindrances facing FM performance and suggested measures to improve facilities' performance within the perception of Nigerian public higher institution buildings. The present state of physical infrastructure development in public higher education institutions calls for urgent attention. This may have contributed to the low ranking of Nigerian universities outside the first 400 universities in the world. Exploring measures to improve facilities' management performance in public higher education institution buildings cannot be over-emphasised. From the study, the chapter proffers measure to improve FM performance within the perception of public higher institution buildings in Nigeria via the following major recommendations:

(i) The managers of higher education institutions (VCs and Rectors) should revamp their facilities management institutional framework. This includes staff training, a viable budget, proactive maintenance plans, match with workable financial support, and an upgrading plan to revitalise obsolete physical buildings to become sustainable and functional in twenty first-century higher education institutions.

(ii) The study recommends more efforts to promote data-capturing technologies in the revitalised or newly constructed buildings to enhance cloud data for developing as-built models for existing and new facilities. Also, digital technologies such as BIM integrated into facilities management enhance sustainability and functionality, especially in physical education facilities.

(iii) Also, innovative digital technologies and tools integrated into facilities management enhance decision-making for the building lifecycle and support for maintenance. The outcome will improve sustainable construction engineering and facilitate 'smart higher education institutions.' Teaching, learning, and research flourish in a smart educational environment.

Acknowledgements and Funding Special thanks to the Auchi Polytechnic Rector (Dr. S. S. Umar) and his Team for the research support. Also, thanks to the participants for providing knowledgeable contributions to enhance the findings of this study. Also, the authors appreciate the comments, suggestions, and recommendations provided by the anonymous reviewers, which honed and strengthened the quality of this manuscript during the blind peer-review process. The research was funded by Universiti Sains Malaysia and the Faculty of Engineering and the Built Environment and CIDB Centre of Excellence (05-35-061890), University of Johannesburg, Johannesburg, South Africa.

Appendix

Dear Participant,

Request for Interview

Effective facility management (FM) has been proved a successful tool to sustain building performance. This technique can enhance the functionality and sustainability of public higher education institution buildings (PHEIBs). How far regarding facilities management performance on Nigerian PHEIBs is yet to receive in-depth studies. Therefore, this research is titled: **Facilities Management Performance in Public Higher Education Institution Buildings in Nigeria.** Specifically, this research will be achieved through the following objectives:

(i) To investigate the perceived hindrances facing FM performance in Nigerian PHEIBs.
(ii) To suggest feasible measures to improve facilities management performance within the perception of public higher institution buildings in Nigeria.

Please note that questions for the interview will be within the study's stated objectives. Responses provided by you will be collated and analysed together with that of other interviewees. It will make up the value and contribution to achieving the success of this work. Information provided will be treated with the greatest secrecy.

Hence, your valuable time and other inputs in answering the questions will be highly cherished.

With regards.

Yours faithfully,

(Research Coordinator)

Basic Questions For The Participants

1. Please, for record purposes, what is the name of your organisation?
2. Please, what is your position in the organisation?
3. Please, how long have you been working?
4. Please, are you knowledgeable regarding facilities management performance and its relationship with Nigeria's public higher education institution buildings?
5. If yes to question 4, what is your take regarding possible challenges implementing facilities management performance on public higher education institution buildings in Nigeria?

6. What are the perceived hindrances facing FM performance in Nigerian public higher education institution buildings?
7. As a stakeholder, do you think digitalisation can play a key role in enhancing facilities management performance in Nigerian public higher education institution buildings?
8. If yes/no, how/why?
9. Please, what role can the key stakeholders play to improve facilities management performance in Nigerian public higher education institution buildings?
10. What feasible ways to enhance facilities management performance on Nigerian public higher education institution buildings?

References

Abdullahi I, Yusoff WZW (2019) Influence of facilities performance on student's satisfaction in northern Nigerian universities: results from interim study. Facilities 37(3/4):168–181. https://doi.org/10.1108/F-08-2017-0088

Aboginije ANT, Thwala WMN (2020) Harnessing fourth industrial revolution (4IR) for improving poor universities infrastructure in developing countries—a review. In: Proceedings of the international conference on industrial engineering and operations management, Held in Dubai, UAE, Mar 10–12

Adegbite JBO (2007) The education reform agenda: challenges for tertiary education administration in Nigeria. A paper presentation at the sixth annual seminar of conference of Registrars of colleges of education in Nigeria (southwest zone), College of education, Ikere-Ekiti, Nigeria

Adewunmi Y, Ajayi C, Ogunba O (2009) Facilities management: factors influencing the role of Nigerian estate surveyors. J Facil Manag 7(3):246–258

Amaratunga A, Baldry D (2000) Assessment of facilities management performance in higher education properties. Facilities 18(7/8):293–301

Amaratunga D, Baldry D (2002) Moving from performance measurement to performance management. Facilities 20(5/6):217–223

Amoo A (2018) Education ministry gets 7.02% of Nigeria's 2019 budget. EDUCELEB.COM. Retrieved from https://educeleb.com/education-ministry-nigerias-2019-budget/#:~:text=The%20Federal%20Ministry%20of%20Education%20has%20been%20allocated,budget%20presentation%20at%20the%20National%20Assembly%20in%20Abuja

Anaelobi ES, Agim EC (2019) TETFund intervention and development of university libraries in South-East, Nigeria. Libr Inf Perspect Res 1:50–58

Anderton D (2016) Science in the city region: establishing Liverpool's life science ecology. Reg Stud Reg Sci 3(1):437–444

Araszkiewicz K (2017) Digital technologies in Facility Management—the state of practice and research challenges. Procedia Eng 196:1034–1042

Asiabaka IP (2008) The need for effective facility management in schools in Nigeria. N Y Sci J 1(2):10–21

Atkin B, Brooks A (2000) Budget 2003: report—chapter 6: delivering high quality public services. In: Total facilities management. Blackwell Science, Oxford, U.K. British Ministry of Finance, 2003. Available at: www.hmtreasury.gov.uk/budget/bud_bud03/budget_report/bud_bud03_repchap6.cfm

Atkin B, Brooks A (2021) Total facility management, 5th edn. Wiley Blackwell, West Sussex, United Kingdom

Atuahene F (2014) Charting higher education development in Ghana: growth, transformations, and challenges. In: The development of higher education in Africa: prospects and challenges international perspectives on education and society, vol 21. Emerald Group Publishing Limited, pp 215–263. https://doi.org/10.1108/S1479-3679(2013)0000021011
Belcher RG (1997) Corporate objectives, facilities, measurement and use: a university model. Proceedings of the RICS Cobra conference, Portsmouth
Benneworth P, Fitjar DR (2019) Contextualizing the role of universities to regional development: introduction to the special issue. Reg Stud Reg Sci 6(1):331–338. https://doi.org/10.1080/21681376.2019.1601593
Bryman A (2016) Social research methods, 5th edn. Oxford University, Oxford, United Kingdom
CABE (2005) Design with distinction: the value of the good building design in higher education. Available at: www.cabe.org.uk. Accessed on 25 Nov 2011
Chitopanich S (2004) Positioning facilities management. Facilities 22(13/14):364–372
Clarke G (1997) Reassessing resource allocation strategies in higher education: methods for analysis'. Int J Educ Manag 11(6):286–292
Creswell WJ, Creswell DJ (2018) Research design: Qualitative, quantitative, and mixed methods approaches, 5th edn. Sage, London, United Kingdom
Ebekozien A (2019) Root cause analysis of demand0supply gap to low-cost housing in Malaysia. PhD thesis submitted to the Universiti Sains Malaysia, Pulau Pinang, Malaysia
Ebekozien A (2020a) Corrupt acts in the Nigerian construction industry: is the ruling party fighting corruption? J Contemp Afr Stud 38(3):348–365. https://doi.org/10.1080/02589001.2020.1758304
Ebekozien A (2020b) A qualitative approach to investigate low-cost housing policy provision in Edo State, Nigeria. Int Plann Stud 1–18. https://doi.org/10.1080/13563475.2020.1779671
Ebekozien A, Samsurijan MS, Aigbavboa C, Awe EO, Amadi GC, Emuchay FE (2022) Unravelling the encumbrances in procurement management of Nigeria's infrastructure development: pitfalls and prospects of projects. Property Manage. https://doi.org/10.1108/PM-11-2021-0103
Elebiju A, Ilesanmi AT (2020) Nigeria: pitfalls, issues and prospects: a perspective on some concessions of public infrastructure in Nigeria. Retrieved from https://www.mondaq.com/nigeria/construction-planning/1016998/pitfalls-issues-and-prospects-a-perspective-on-some-concessions-of-public-infrastructure-in-nigeria-
Enefola FO (2016) Infrastructural development in Nigeria: a panacea for transforming educational sector for sustainable national development. Int J Capacity Building Educ Manage 3(1):13–19
Fenker M (2004) Organisation changes, representations and facilities. In: Alexander K (ed) Facilities management: innovation and performance. Taylor Francis
Gbadegesin JT, Aluko BT (2014) Public private partnerships/private finance initiatives for financing infrastructure in public tertiary institutions in Nigeria. Built Environ Proj Asset Manage 4(2):199–215
Ibrahim FS, Ebekozien A, Khan P, Aigbedion M, Ogbaini IF, Amadi G (2022) Appraising fourth industrial revolution technologies' role in the construction sector: how prepared is the construction consultants? Facilities. https://doi.org/10.1108/F-09-2021-0086
Ikediashi DI, Ogunlana SO, Bowles G, Mbamali I (2012) Outsourcing of facilities management services in Nigeria's public universities. In: Laryea S, Agyepong SA, Leininger R, Hughes W (eds) Proceeding 4th West Africa built environment research (WABER) conference, 24–26 July 2012, Abuja, pp 725–735
International Facility Management Institution-IFMA (2006) What is FM? Available at: www.ifma.org/what_is_fm/index.cfm
Jaafar M, Ebekozien A, Mohamad D (2021) Community participation in environmental sustainability: a case study of proposed Penang Hill Biosphere Reserve, Malaysia. J Facil Manage. https://doi.org/10.1108/JFM-03-2021-0033
Johnston A (2007) Final report of international action research project based on research carried out by Dr Andy Johnston. Seconded to the OECD from forum for the future, in Oct 2006–Mar 2007 (Oct 2006), pp 1–62

Kaplan RS, Norton DP (1996) The balanced score card. Harvard Business School Press, Boston, MA

Lavy S (2008) Facility management practices in higher education buildings. J Facil Manag 6(4):303–315

Marmolejo F (2007) Higher education facilities: issues and trends. PEB Exchange 2007/1 ISSN 1609-7548

Moore M, Finch E (2004) Facilities management in South East Asia. Facilities 22(9/10):259–270

National Universities Commission (NUC) (2006) Ban on admission into more programmes with denied accreditation. Monday Memo 4(18)

Needs Assessment of Nigerian Public Universities (2012) CNANU NEEDS assessment report presentation to NEC. Available at: www.fuotuoke.edu.ng/.../CNANUNEEDSAssessmentReportPresented

Odediran SJ, Gbadegesin JT, Babalola MO (2015) Facilities management practices in the Nigerian public universities. J Facil Manag 13(1):5–26. https://doi.org/10.1108/JFM-11-2013-0058

Ogungbile JA, Oke EA (2015) Assessment of facility management practices in public and private buildings in Akure and Ibadan cities, south-western Nigeria. J Facil Manag 13(4):366–390

Olufemi F (2020) Buhari's 2021 budget share for education is Nigeria's lowest in 10 years. Premium Times. Retrieved from https://www.premiumtimesng.com/news/headlines/422829-buharis-2021-budget-share-for-education-is-nigerias-lowest-in-10-years.html

Omotayo ST, Boateng P, Osobajo O, Oke A, Obi IL (2020) Systems thinking and CMM for continuous improvement in the construction industry. Int J Prod Performance Manage 69(2):271–296. https://doi.org/10.1108/IJPPM-11-2018-0417

Preiser WFE (1995) Post-occupancy evaluation: how to make buildings work better. Facilities 13(11):19–28

Price I (2003) Facility management as an emerging discipline. In: Best R, Langston C, De Valence G (eds) Workplace strategies and facilities management building in value. Butterworth-Heinemann, Oxford

Rose R, Cain DA, Dempsey JJ, Schneider R (2007) Buildings ... the gift that keep on taking-a framework for integrated decision making. APPA and the Center for Facilities Research, Alexandria, VA

Saldana J (2015) The coding manual for qualitative researchers, 3rd revised edn. Sage, London, United Kingdom

Sani AM (1998) Emergent trends in facility management. Paper presented at NIESV, 28th annual conference, Kano

Saunders MNK, Lewis P, Thornhill A (2012) Research methods for business students, 6th edn. Pearson Education, Harlow, England

Sekaran U, Bougie R (2016) Research method for business: a skill building approach, 7th edn. Wiley, West Sussex, UK

TETFund (2011) About TETFund. Retrieved from http://www.tetfund.gov.ng

Then DSS (1999) An integrated resource management view of facilities management. Facilities 17(12/13):462–469

UNESCO (2006) World enrollment of tertiary education in world bank countries. Available at: www.upo.unesco.org

Wentworth L, Makokera GC (2015) Private sector participation in infrastructure for development. S Afr J Int Aff 22(3):325–341. https://doi.org/10.1080/10220461.2015.1081568

Wong JKW, Ge J, He SX (2018) Digitisation in facilities management: a literature review and future research directions. Autom Constr 92:312–326

Yoon JH, Cha HS (2017) An empirical study of facility management performance on commercial office building in South Korea. In: Civil engineering and urban planning: proceedings of the 5th international conference on civil engineering and urban planning (CEUP2016). pp 151–159

Chapter 5
Accessibility of Public Conveniences by Students in Higher Education Institutions

A. O. Adewusi and J. A. Oguntokun

Abstract Public conveniences are important to everyone, irrespective of age, class, ethnic origin, gender, mental ability, or physical ability. Access to it or otherwise serves significant implications on the total learning environment. This paper discusses the accessibility of students to public conveniences in a major academic institution in South West Nigeria. This was a cross-sectional descriptive study conducted among 233 students in the institution. Both quantitative and qualitative methods were used in collecting data for the study. Qualitative data was collected using in-depth interviews, observation, and physical inspection of sanitary facilities in the faculties, while a structured self-administered open-ended questionnaire was used to gather data on students' assessment of sanitary facilities in the selected faculties. Descriptive statistics and weighted mean scores were used in analysing the data using SPSS version 21. The analysis reveals among other that the state of sanitary facilities across the faculties and hall of residences are in a deplorable state, the student to toilet ratio is as high as 1:195 as against the standard requirements recommended by National University Commission (NUC), the available ones are poorly maintained with a high number of toilet facilities under lock and key resulting in students practising unhygienic practices of open defecation. The findings also reveal that lighting, ventilation, adequate water supply, good management and maintenance, privacy and security, availability of toilet supplies, location, availability of disabled toilets, aesthetics, and segregated toilets with signages are the main factors determining utilisation of public toilet facilities It was recommended that toilet facilities should be made available and accessible for all, irrespective of ability or disability to foster a healthy learning environment.

Keywords Disabled · Toilets · Accessibility · Public facility

A. O. Adewusi · J. A. Oguntokun (✉)
Department of Estate Management, School of Environmental Technology, Federal University of Technology, Akure, Nigeria

A. O. Adewusi
e-mail: aoadewusi@futa.edu.ng

S. P. Akinbogun et al. (eds.), *Facility Management practices*,
https://doi.org/10.1007/978-3-031-65321-6_5

5.1 Introduction

Access to and efficient use of safe sanitation facilities is an essential part of public health, this underscores the need for the provision of decent toilet facilities for all. Despite their importance, public toilets present a challenge for every society in Nigeria. Anyone who has needed a public toilet has invariably acquired the knowledge that one nearby is priceless, and such facilities have been consistently poorly managed (Drangert and Greed 2010). Kolsky (2006) defined public conveniences as toilets that allow people to meet their sanitary needs in public places such as markets, and transport centres. Pathak (1995) also defined it as a critical link between order and disorder and between good and bad environments as the need to urinate and defecate is irrepressible. Although we use these toilets every day, at home, school, work, restaurants, shopping malls, and places of worship, we seldom talk about them. The silence around the issue of toilets and sanitation has deadly consequences. Greed (2006) reported that a lack of toilet facilities at the right time in the right place contributes to dirty streets that are unsanitary, and unpleasant can spread infection. A study conducted by Officer and Posarac, (2011), revealed that globally, there are approximately 2.6–3 billion people without proper sanitation. The worst situation is in Southern Asia and sub-Saharan Africa, where the demand for sanitation is rising due to the growing population, especially in urban areas. Historically, government Institutions in developing countries have lacked basic infrastructure and facilities such as accessible and clean toilets, covered classrooms, electricity, sewage systems, and clean drinking water, and this lack of infrastructure may be responsible for high student absenteeism rates (Jasper et al. 2012) and high dropout rates (Adukia 2017) mainly because of diseases and others.

According to Agbo et al. (2012), facilities for academic and non-academic activities need to be properly put in place to provide an optimal sanitary environment that is safe and conducive for the physical, mental, and emotional health of the student to achieve maximum benefits from educational programmes. Deegener et al. (2009) reported that Proper school sanitation is a question of students' health and dignity. However, in many institutions, little attention is paid to school sanitation not only because of lacking financial resources but also due to a lack of awareness of how important school sanitation is. Students in public institutions face the daily challenge of finding a bush, abandoned buildings, or empty lots where they can urinate and defecate in relative privacy. To enable learners with impairments to apply for higher education, institutions must have adequate enabling mechanisms in place to provide access and additional support to students with impairments where needed (Department of Education 1997). The condition of inaccessible toilets can force disabled people into unhygienic and inappropriate practices (Ahsan 2002; Sachelo 2002). For students using a wheelchair, toilet access can be problematic. The provision of equal access to facilities and services to all people with different characteristics should be top on the agenda (Imrie and Kumar 1998). Yet in most cities and campuses, physically challenged individuals cannot easily access facilities of the built environment (Bromley et al. 2007; Chin 2011; Sapey et al. 2005). Research by Fadairo and Adedeji (2012) reveals that the hygienic and sanitary conditions of many rural school toilets in Nigeria range from bad to terrible. Nigerian tertiary institutions have suffered serious neglect and left to cater for their own environmental sanitation needs in terms of public toilet provision, planning, and management. This is even though tertiary institutions in Nigeria accommodate the largest majority of enlightened people in the country. The Nigerian tertiary institutions are characterised by several drawbacks to a sustainable neat environment. These include inadequate toilet facilities in public buildings, lack of constant water supply to the available toilets, and dwindling power supply among others. These have translated into several problems thereby exposing university communities into a series of health risks (Fadairo and Adedeji 2012).

Although the National University Commission NUC (2006), recommends one toilet to ten students (1:10) as the standard requirement in every higher institution in the country, it is evident in most public institutions that the toilets available are not adequate, the available ones are not well maintained with a high number of toilet facilities under lock and key, and therefore not accessible for students use. Additionally, most of these toilet facilities also fall below the standard specification in terms of design and where built conveniences are available, access needs of disabled people are rarely considered nor were they modified to accommodate this disadvantaged members of the society (Sachelo 2002; Eleweke 1999). Research by Kolo et al. (2017) revealed that, provision of toilet facilities is considered a privilege rather than a necessity by most school authorities; which may affect school programmes failing in their roles to address health issues of the students but rather concentrating on the academic pursuits of these students. Therefore, this research attempts to bridge a gap, by exploring the current situation of toilet facilities in a major academic institution in the Southwestern part of Nigeria, and also investigate the factors that prompt students to either use the toilet or practice open defecation in the study area.

5.1.1 Condition of Toilet Facilities in Public Institutions

According To toilet association of Singapore (2003), a well-designed public toilet can be defined as one that is (a) Clean and dry (b) Well ventilated (c) Easy to maintain (d) Carefully planned layout (e) Friendly to persons with disabilities and special needs. Schools partly determine students' health and well-being by providing a healthy or unhealthy environment. Although water and sanitation facilities in schools are increasingly recognized as fundamental for promoting good hygiene behaviour and children's well-being, many schools have very poor sanitation facilities. For some, the conditions vary from inappropriate and inadequate sanitary facilities to the outright lack of latrines and safe water for drinking and hygiene (UNICEF 2007).

Koopman (1978) examined diarrhea and school toilet hygiene in Cali, Colombia. The prevalences of diarrhea, vomiting, common cold, and head lice in schoolchildren were measured In relation to classroom size and to the condition of the school toilets. The study found that unhygienic toilet conditions were related to diarrhea, and it was estimated that if all schools could reach the modest level of hygiene, diarrhea would be reduced by 44% and vomiting by 34% Jasper et al. (2012) reviewed literatures on the impacts of water and sanitation inadequacies in the academic environment. The review concluded that studies document higher rates of infectious, gastrointestinal, neuro-cognitive and psychological illnesses where school children were exposed to inadequate water and sanitation facilities. Kanayochukwu et al. (2020) examined the conditions of educational facilities in public schools in selected Local Government Areas in Kaduna State, Nigeria. Physical inventories were carried out using the Educational Facilities Assessment Form (EFAF). Multistage sampling technique was used to select 9 Local Government Areas (LGAs). Data were analyzed using Geographically Weighted Regression (GWR) 4.10 and Spearman Rank Correlation. The findings revealed that 88% of the schools do not have water supply while 22% have no roof, and 11.4% have power supply. It discovered that 32.9% of the schools were mud buildings while 4.2% has no structure and less than 24% has toilet facilities.

Tiswin et al. (2019), assessed the types, condition and functionality of water, sanitation and hygiene facilities in public primary schools in the Zabzugu District of the Northern Region of Ghana. A cross-sectional descriptive survey was carried out in February 2015. Twenty-five schools were randomly selected for the study. The study found that the available handwashing facilities were underutilised, there was a shortfall of WASH infrastructure in most of the public primary schools in the Zabzugu, including disability-friendly toilets, urinals and menstrual hygiene management facilities. Wada et al. (2020) examined the classification of Sanitation Services and Students' Sanitation Practices among Schools in Lagos, Nigeria. A cross-sectional study utilized a 4-stage sampling technique to select 386 students from schools in Badagry, Lagos. Pre-tested questionnaire and observational checklists were used to obtain data. Data were analyzed using descriptive statistics and logistic regression at 5% level of significance. Findings show that t all the schools had improved sanitation facilities, while 37% of the available toilet compartments were inaccessible to the students. Majority (85%) of the facilities provided limited service, while 15% provided basic service. The student to toilet ratio for the public school girls and boys were 3191:1 and 642:1 respectively, while the private school had a ratio of 257:1 and 289:1 for girls and boys respectively.

5.1.2 Factors that Influence Students' Decision to Use Public Toilets or Practice Open Defecation

Ayee and Crook (2003), opined that without public toilet facilities, people will be compelled to defecate in the open. Public toilet facilities have evolved to become vital components in creating sustainable, accessible, and inclusive cities (Greed 2006). Jones et al. (2012) defined Open defecation as the practice of defecating outside and in public, waterways, as a result of no access to toilets, latrines. The decision of people to use public conveniences is based on the following identified factors.

(a) **Availability of water**

Although the importance of water and sanitary facilities for learning institutions is acknowledged, in practice the sanitary situation in many schools is deplorable. The direct consequences of lack of access to safe drinking water and sanitation services are enormous. The availability of water supply will encourage students to practice safe hygiene, however the lack of water supply will result in the spread of diseases and toilet infections especially among female students. Research by Kabir et al. (2021) examined the factors influencing sanitation and hygiene practices among students in a public university in Bangladesh. Semi-structured interview was conducted to elicit information from respondents in the study area Findings show that despite having reasonable awareness and knowledge, the sanitation and hygiene practices of the students were remarkably low. A broad array of interconnected factors influenced sanitation and hygiene behavior, as well as each other. Individual factors (gender, awareness, perception, and sense of health benefits), contextual factors (lack of cleanliness and maintenance, and the supply of sanitary products), socio-behavioural factors (norms, peer influence), and factors related to university infrastructure (shortage of female toilets, lack of monitoring and supervision of cleaning activities) emerged as the underpinning factors that determined the sanitation and hygiene behavior of the university going-population. Water shortage in institutions has been identified as the one of the major causes of open defecation, however, research by Andavar et al. (2020) clearly indicates that the requirements of the water to use for flushing, cleaning, and washing toilets in the administrative buildings and classrooms can be satisfied by using rainwater as an alternative to tap water.

(b) **Management and maintenance of school conveniences**

Buxton et al. (2019) examined the impact of a school toilet operation and management intervention in the Philippines on toilet usability and student and teacher satisfaction. A non-blinded cluster randomised controlled trial. The study compared improvements in usability and cleanliness of school toilets among those schools receiving a low-cost, replicable intervention. Toilet usability was measured based on Sustainable Development Goal indicators related to school sanitation defined by the UNICEF/WHO Joint Monitoring Programme for Water, Sanitation and Hygiene. 20 eligible schools in the Batangas region of the Philippines were randomly selected and allocated to either control or intervention group. Findings show that non-classroom toilets were 48% more likely to meet quality benchmarks in intervention schools. Drangert and Greed (2010) reported that Cleanliness has significant implications in public toilet management and patronage as people are easily put off by what they see. People are appalled when they see faecal stains or excreta, nauseating odour of urine and used tissue paper left unattended to by previous user(s) designing and building of a public toilet block is half the battle as the other half battle is to manage, clean, repair, maintain and operate it over many years.

All this needs to be backed up by efficient and regular cleaning regimes (Greed 2006). This is important because not maintaining and cleaning public toilets results in a proliferation of toilet-related diseases and medical conditions (Greed 2005). A survey of sanitation facilities in schools in Bloomsbury, London by Jewekes and O'connor (1990) found that a large number of the toilets were not kept adequately clean. The authors described this situation as potential for the spread of infectious diseases. Cleaning staff should be seen as the front-line troops in disease control as dirty toilets attract flies, lice, rats, mice, fleas, ticks and so forth which are vectors for spreading disease Greed (2006).

(c) **Disabled toilets**

Access to toilets is also essential for disabled students. Pierce (1998) found that one of the commonest complaints of physically disabled people was the inaccessibility of the environment caused by architectural barriers. Inaccessibility can force disabled people into unhygienic and inappropriate practices (Ahsan 2002; Sachelo 2002). Kitchin and Law (2001) reveal three critical themes in relation to the provision of public toilets for disabled people: (i) a lack of accessible toilets because of vandalism, and poor and limited provision; (ii) poor toilet design (e.g., steps, inadequate door size); and (iii) misuse of toilets (e.g., they are sometimes used as storage cupboards and/or are kept locked). As a result of these, disabled people exhibit constrained patterns of spatial behaviour and daily routines. Squat toilet facilities are common. African toilets have no sitting pans, which means people with impaired lower limbs cannot use them (Sachelo 2002). Kiwanuka (2002) reports that 'some latrines are a long distance from their hostel or home or class, and many do not have ramps. Some students have difficulty opening the doors to enter, and closing while inside—the door locks are sometimes too high for them, and space inside is limited to freely turn and transfer. Cleaning after use and washing hands is also hard, the taps seem to be too high. For those crawling they find the floor too dirty, yet they use bare hands at times when crawling.' Where a water source is not close by, the user may face the problem of how to carry water to the latrine for washing (Mulholland et al. 1998).

Serlin (2017) investigated the discourse about equality in public restrooms in the US and highlighted that while public toilets became common throughout the nineteenth and twentieth centuries, they lacked accessibility and usability for disabled people. Due to inappropriate and non-existent facilities, the disabled were excluded from the public sphere. Again, this situation highlights the need for designers to think about this privileged-access issue and how to make public toilets accessible to, and usable for, all. Because "architects, planners, and builders until very recently have (had) no obligation to provide accessible toilets in public spaces and buildings, consequently few were installed" (Kitchin and Law 2001).

(d) **Design and Location**

The provision of public toilets is not only a matter of land use, but also an essential design and planning concern to make cities more accessible, inclusive, and convenient for all members of society. Michael and Olamiju (2015) opined that a lack of available and appropriately located public toilet facilities at the right time during the day and night encourages street fouling seriously impairing the quality of place and quality of life for local people. Greed (2005), opines that toilets should be proudly placed out in the open and not hidden but thoughtfully designed. Toilets should be located in central public thoroughfares, squares, in open well-lit areas, and people should be proud of them as an important townscape statement in their own right (Greed and Roberts 2014). Public toilets that are badly designed, badly maintained, and poorly located generate a sense of neglect, attracting vandalism, anti-social behaviour, and social disorder (Oyinloye and Oluwadare 2015).

(e) **Adequate ventilation**

Proper ventilation of a public toilet is one of the highest priorities in the design of toilets. An ineffective ventilation system can make a public toilet unbearable, even if it is well designed. An effective ventilation system ensures that vitiated air is quickly extracted, and helps to avoid dampness and subsequent growth of mould on floors and walls. Natural ventilation should be achieved through windows, doors, louvers, or other openings to the outdoors. Chung et al. (2019) researched the restrooms in public schools (including senior high schools, junior high schools, and elementary schools) in Taiwan. They analyzed the influence of different numbers of awning windows and their installation position on the indoor airflow patterns and air contaminant distributions in a restroom. The results showed that an adequate number of awning windows and appropriate installation positions helped achieve the best ventilation effect. They also offered advice on where awning windows should be installed. Lee and Tham (2021) investigated 61 public toilets to identify the causes and locations of biological contaminated sources. Airborne and surface bacterial contamination, carbon dioxide concentration, and surface ammonia levels were measured. Findings show that both bacterial contamination and CO_2 are higher in non-ventilated toilets compared to their ventilated counterparts.

(f) **Security and privacy**

Privacy and safety in public toilet facilities in urban environments are a paradox as providing discretion could also induce fear and isolation. Therefore, even though the location of public toilets is a sensitive matter, planners rarely consider public toilets as a land-use matter. However, deciding on public toilet locations requires a delicate balance between providing natural discretion but ensuring the place is open enough to assuage the patrons' fears of isolation (Greed 2006). Ellisa and Luana (2022) conducted a spatial study on women's public toilets, focusing on how the situational and spatial condition of the toilets influence and challenge women to utilize, avoid, or delay using public toilets. The study was conducted in Kotatua Jakarta. The research method was experience-based user interviews combined with spatial and locational studies, which focused on the surveys and mapping of attribute descriptions and observation behavior. It was found that as well as suitable social and cultural environments, to avoid crime and the fear of crime, female public facilities should have crime prevention mechanisms on the physical nodes (activity), the paths (route), and the edges (the boundary of awareness). Hayat (2017), reported that the lack of basic infrastructure discourages households in Pakistan from enrolling their children in schools and this reluctance is more pronounced in the case of girls, primarily due to a concern for their safety, as well as to protect them from sexual harassment and health problems.

(g) **Toilet supplies and facilities**

A study of primary school children in Malaysia reiterated that improved water supply, toilet facilities, and sanitation have lowered the prevalence of waterborne diseases in the study area. Simple hand washing with soap and water reduces Shigella and other diarrhoea transmission by 35% (Kotloff et al. 1999; Khan 1982; White et al. 2003). Availability of toiletries such as soap dispensers, sensor flush, litter bins, electronic hand dryers/paper towel dispensers, toilet seat sanitizers, full-length mirrors, air fresheners, WC/Urinals sanitizers, and coat hooks are essential school toilets. In a study of the effect of hand hygiene on illness rate among students in university residence halls, White et al. (2001) confirmed that hand hygiene using alcohol gel sanitizer decreases the incidence of upper respiratory illness among students living in university residence halls from 14 to 39.9% and total improve illness to 20% and absenteeism.

(h) **Segregated toilets with signages**

Research by Save the Children (2009), Lidonde (2005) reported that lack of adequate, well fitted, and segregated sanitation facilities at schools discourages a number of Nigerian girls from attending full time education system, which consequently affect their academic performance and perpetuate gender inequity According to the Statement of the Committee on the Right to Sanitation (45th session, E/C.12/2010/1) of the United Nations Committee on Social, Economic, and Cultural Right, "girls do not go to school in many parts of the world for lack of toilets, or lack of separate toilets for them" (UN 2010). Accordingly, most schools are built without separate toilets for the male and female, for example Adukia (2014) formed separate estimates by gender and age, and found that the availability of segregated toilets is especially largely correlated with improved learning outcomes and school attendance for pre-pubescent female students.

Greed (2004), Drangert and Greed (2010) opine that the use of signage is very important as it depicts how to find available public conveniences. Signage facilitates gives direction to sanitary facilities and category of users (Oyinloye and Oluwadare 2015). Restroom association of Singapore (2013) suggested that Clear signage should be designated for each gender of required public toilet facilities. The location of the signs should be near the entrance to each toilet facility and clearly displayed at noticeable locations in main traffic passageways to direct the public to the toilets.

(i) **Lighting**

A well-designed lighting system will save electrical energy and improve the appearance of the toilet. Poorly designed fixtures with discoloured diffusers go a long way to make a toilet dingy. Dark and shadowy, off-coloured lighting can create the impression that a toilet is not clean. Natural lighting can be used to help create a softer, friendlier environment. Harsh lighting can create a cold and unwelcoming air while being inappropriate for the tasks being performed. It can also highlight hard-to-clean areas. Restroom association of Singapore (2013) opined that all public toilets should be provided with warm-colour lighting for general lighting as well as down lights above the wash basin/mirror. The provision of emergency illumination devices is also necessary to illuminate the path of egress for occupants to exit the toilet safely.

(j) **Aesthetics**

The use of artworks, colourful tiles and bright colour wall finishes can be used to create ambience in the toilets, to make it more welcoming and comfortable for students use. Toilet association of Singapore (2003) suggested that the ambience of public toilets can be enhanced further by: (a) Introducing plants which can be easily maintained inside the toilets as well as surrounding the public toilets. (b) Placing of wall pictures and illuminated with delicate lighting. The pictures or wallpapers should be waterproof (e.g. made of impervious material). (c) Placing of ornaments or sculptures at the 'dead' corners of the toilets.

5.2 Methodolgy

5.2.1 Research Design

This study was designed to examine the accessibility of public convenience to students in a major academic institution in Southwest Nigeria. This type of investigation can be regarded as both descriptive and correlational research. The participant observation was only done to ascertain the nature of the toilet facilities. The study was conducted in selected faculties in the institution. This study covers both the able body and disabled students in the university (i.e. students with disabilities in the lower limbs, using crutches, wheelchairs, and walking sticks). The toilets under study are the toilets within each Faculty. These includes toilets within Faculty of Agriculture, Faculty of Engineering, Faculty of Sciences, Faculty of Mineral Science, Faculty of Built Environment, and the Faculty of Management. These toilets are designed with the aim of providing a sanitary facility for defecating and urinating for students in these Faculties. There is no user barrier as students, workers and passers-by are all privileged to use these toilets.

5.2.2 Population and Sample Size

The research population for this study include both the able bodied and the physically challenged students in the University. The total population of students in the school for the 2017/2018 academic session was 17,124. There are a total of 8 Faculties which are; Faculty of Agriculture 2382, Faculty of Engineering 2980, Faculty of Sciences

2606, Faculty of Mineral Science 2150, Faculty of Built Environment 3555, the Faculty of Management 2345, Faculty of Computing 395, and Faculty of Health 711. Data on disabled students was obtained from the student affairs. A total of 14 disabled students are currently enrolled in the school across all levels.

5.2.3 Sampling Technique

Purposive sampling method was adopted for this study. Agyedu et al. (2010) suggested that in purposive sampling, the researcher uses his or her judgment about which respondents to choose, and picks only who best meet the purpose of the study. A purposive sampling technique was used to select Six (6) faculties viz; Faculty of Agriculture, Faculty of Engineering, Faculty of Sciences, Faculty of Mineral Science, Faculty of Built Environment, and the Faculty of Management. This is because these Faculties have the higher number of students. The proportion of disabled students are few (14 people), as a result, census method was adopted.

5.2.4 Data Collection Instruments

Each of the sampled faculties has a population of over one thousand (1000) students. A total of fifty (50) questionnaires representing an estimated 5% of the school population was administered in each of the schools in equal proportion to both male and female students, bringing the total number of respondents to three hundred (300). However, 233 questionnaires were recovered, which forms the basis of the data analysis.

5.3 Demographic Characteristics of Respondents

The demographic characteristics of the respondents considered in this study includes their Gender (Table 5.1) and Disability (Fig. 5.1), this is due to the fact that the study sample comprises of both able and disabled students (4 disabled students, i.e. 3 male and a female.). Table 5.2 shows the Academic levels of the respondents out of which 27.89% are in 100 level, 23.17% are in 200 level, 26.60% are in 300 level, 17.16% are in 500 level, while 5.50% are postgraduate students.

5.3.1 Results and Discussion

1. ***The prevailing condition of the toilet facilities in the school***

Table 5.1 Gender of respondent

Gender	Frequency	Percentage (%)
Male	121	52
Female	112	48
Total	**233**	**100**

Authors field survey 2018

Fig. 5.1 Distribution of respondent in terms of disability. Authors field survey 2018

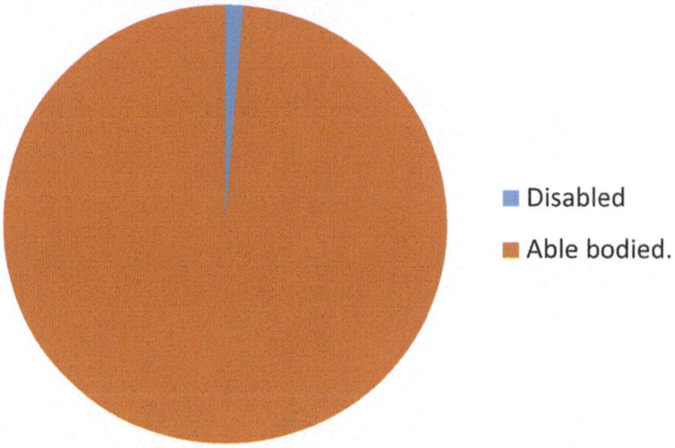

Table 5.2 Academic level of respondents

Level	Frequency	Percentage
100	65	27.89
200	54	23.17
300	62	26.60
500	40	17.16
PG	12	5.15
Total	**233**	**100**

Authors field survey 2018

Table 5.3 shows the comprehensive list of the basic important facilities expected in a Public University setting which includes; Doors and windows for security and ventilation, lighting, wash hand basin, water, soap/toiletries/sanitizers/disinfectants, and facilities for disabled. The result shows that all the faculties have good doors and windows, good wash hand basins, however, there is no functional lighting and inadequate water supply in most of the schools, Soaps, sanitizers and disinfectants are lacking in all the toilets in the faculties, and there is no accessible toilet provision for the disabled students in any of the faculties. All of the toilets in the faculties are somehow accessible for students use, except for Faculty of Management in which none of the toilets are opened for students use.

Table 5.4 shows the total number of toilets and students in each school in relation to the toilets available. To determine the student toilet-ratio, the total number of

Table 5.3 Condition of school conveniences

Faculties	Doors and windows		Lighting		Wash hand basin		Water		Soap, toiletries and disinfectants		Toilet facilities for disabled		Accessibility of facilities		
	G	B	G	B	G	B	G	B	A	NA	A	NA	A	SA	NA
FOS	*			*	*			*		*		*		*	
FOBE	*			*	*			*		*		*		*	
FOMS	*		*		*		*			*		*		*	
FOE	*			*	*			*		*		*		*	
FOM	*		*		*			*		*		*			*
FOA	*			*	*		*			*		*		*	

Authors field survey 2018

students in each school was divided by the corresponding total number of toilet facilities available. Table 5.4 shows that FOM and FOS were having the highest student toilet ratio with 2345 and 3555 total number of students and a toilet ratio of 1:195 and 1:148 respectively. While FOMS had the lowest student toilet ratio with a total of 2150 students and a toilet ratio of 1:27, FOA, FOE, and FOBE on the other hand have a student population of 2382, 2989 and 2606 respectively with a toilet ratio of 1:60, 1:60, and 1:44 respectively.

2. ***Factors that influence students decisions to use public toilets or practice open defecation***

Table 5.5 below shows the ranking to ascertain factors that influence students decisions to use conveniences available in the faculties or practise open defecation, with the likerts scale of (MS = Most significant, S = Significant, UD = Undecided, LS = Least significant, NS = Not significant) Therefore, it appeared that from the mean item scores (MIS), Lighting has the highest MIS of 4.463, followed by Ventilation having a mean score of 4.364, Water ranked 3rd with a score of 4.339, while maintenance/Management ranked 4th having a mean score of 4.334, Privacy and Security was ranked 5th with a mean score of 4.309, while Toilet supplies and Location both rank 6th with a mean score of 3.841, Accessible toilets ranked 7th with a mean score of 3.506. Aesthetics ranked 8th, with a mean score of 3.424. Segregated toilets with signages ranked 9th, with a mean score of 3.321.

Table 5.4 Number of toilets in the school

School	Total no of students	Male toilets	Female toilets	Total no of toilets	Student to toilet ratio
FAT	2382	20	20	40	1:60
FOE	2980	20	20	40	1:60
FOBE	2606	**	**	18	1:144
FOMS	2150	40	40	80	1:27
FOS	3555	**	**	24	1:148
FOM	2345	**	**	12	1:195

Authors field survey 2018

Table 5.5 Factors that influence students decision to use school toilets or practise open defecation

S/N	Factors	Ms	S	UD	LS	NS	Mean	Rank
1	Lighting	140(60)	79(33.9)	2(0.85)	6(2.5)	6(2.5)	4.463	1st
2	Ventilation	131(56)	81(34.7)	4(1.7)	9(3.8)	8(3.4)	4.364	2nd
3	Water	114(48.9)	98(42.0)	10(4.29)	8(3.40)	3(1.28)	4.339	3rd
4	Management and maintenance	125(53.6)	79(33.9)	15(6.40)	10(4.2)	4(1.70)	4.334	4th
5	Privacy and security	128(54.9)	81(34.7)	7(3.0)	2(0.85)	15(6.4)	4.309	5th
6	Toilet supplies	97(41.6)	74(31.7)	3(1.28)	46(19.7)	13(5.5)	3.841	6th
7	Location	93(39.9)	72(30.9)	12(5.1)	50(21.4)	6(2.5)	3.841	6th
8	Accessible toilets	42(18.0)	64(27.4)	86(36.9)	17(7.2)	24(10.3)	3.506	7th
9	Aesthetics	58 (24.8)	92 (39)	10 (4.2)	37(15.8)	36(15.4)	3.424	8th
10	Segregated toilets with Signages	61(26.0)	52(22.3)	45(19.0)	51(21.8)	24(10.3)	3.321	9th

Authors field survey 2018

5.4 Conclusion

Based on the findings of this study, it was concluded that basic toilet facilities exist in all the schools, but are in poor condition and not in sufficient quantity, the available ones are dirty and unhealthy for users. This trend explains why people use open spaces when they need to relieve themselves (Drewko 2007), which can predispose students to diseases outbreaks. Majority of public universities in Nigeria lack the facilities to support integrated and inclusive education for the disabled, this is evident in the findings of the study which shows that none of the toilets facilities supports use by students with disabilities. When public toilets are well designed, managed, and maintained, they can yield healthful and accessible environments for all people and embody a vision of positive and equal human participation in urban spaces.

References

Adukia A (2017) Sanitation and education. Am Econ J Appl Econ 9(2):23–59

Agbo HA, Envuladu EA, Zoakah AI (2012) A comparative assessment of the epidemiology of malnutrition among pupils in public and private primary school in Jos North local government area of Plateau State. Jos J Med 6(2):34–38

Agyedu GO, Donkor F, Obeng S (2010) Teach yourself research methods, University press, KNUST, Kumasi, Ghana, pp 95–103
Ahsan MT (2002) Sanitation for the people with physical disabilities in Bangladesh. Personal communication (e-mail, 29.6. 02). Centre for Services and Information on Disability, Dhaka, Bangladesh
Andavar V, Ali BJ, Ali SA (2020) Rainwater for water scarcity management: an experience of Woldia university (Ethiopia). J Bus Econ Environ Stud 10(4):29–34
Ayee JR, Crook R (2003) "Toilet wars": urban sanitation services and the politics of public-private partnerships in Ghana, vol 213. Institute of Development Studies
Bromley RD, Matthews DL, Thomas CJ (2007) City centre accessibility for wheelchair users: the consumer perspective and the planning implications. Cities 24(3):229–241
Buxton H, Dimaisip-Nabuab J, Duijster D, Monse B, Benzian H, Dreibelbis R (2019) The impact of an operation and management intervention on toilet usability in schools in the Philippines: a cluster randomised controlled trial. BMC Public Health 19:1–11
Chin SL (2011) Building and infra-structural design and facilities for the disabled. Doctoral dissertation, UTAR
Chung SC, Lin YP, Yang C, Lai CM (2019) Natural ventilation effectiveness of awning windows in restrooms in K-12 public schools. Energies 12(12):2414
Deegener S, Wendland C, Samwel A, Samwel M (2009) Sustainable and safe school sanitation. In: How to provide hygienic and affordable sanitation in areas without a functioning wastewater system
Department of Education (1997) Quality education for all: overcoming barriers to learning and development. Report of the NCSNET/NCESS
Drangert JO, Greed C (2010) Public toilets: sustainable sanitation for the 21st century. Linkoping University, Sweden
Eleweke CJ (1999) The need for mandatory legislations to enhance services to people with disabilities in Nigeria. Disabil Soc 14(2):227–237
Ellisa E, Luana L (2022) Female restrooms in the tourist destination: how the socio-spatial conditions of public toilets influence women's perception of safety. J Asian Architect Build Eng 21(4):1624–1639
Fadairo G, Adedeji YMD (2012) The menace of public toilets in institutional housing: case studies of selected universities in South-West Nigeria. WIT Trans Ecol Environ 160:125–135
Greed C (2004, June) Public toilets: the need for compulsory provision. In: Proceedings of the institution of civil engineers-municipal engineer, vol 157, no 2. Thomas Telford Ltd, pp 77–85
Greed C (2005) Taking stock: an overview of toilet provision and standards. World Toilet Conference, Belfast, p 14
Greed C (2006) The role of the public toilet: pathogen transmitter or health facilitator? Build Serv Eng Res Technol 27(2):127–139
Greed C, Roberts M (2014) Introducing urban design: interventions and responses. Routledge
Hayat FA (2017) The relationship between access to toilets and school enrollment in Pakistan. Doctoral dissertation, Georgetown University
Imrie R, Kumar M (1998) Focusing on disability and access in the built environment. Disabil Soc 13(3):357–374
Jasper C, Le TT, Bartram J (2012) Water and sanitation in schools: a systematic review of the health and educational outcomes. Int J Environ Res Public Health 9(8):2772–2787
Jewkes RK, O'Connor BH (1990) Crisis in our schools: survey of sanitation facilities in schools in Bloomsbury health district. BMJ 301(6760):1085–1087
Jones H, Fisher J, Reed R (2012, September) Water and sanitation for all in low-income countries. In: Proceedings of the institution of civil engineers-municipal engineer, vol 165, no 3. Thomas Telford Ltd, pp 167–174
Kabir A, Roy S, Begum K, Kabir AH, Miah MS (2021) Factors influencing sanitation and hygiene practices among students in a public university in Bangladesh. PLoS ONE 16(9):e0257663

Kanayochukwu EC, Dogo B, Oluwole OA, Oluwafemi AT, Ajibuah J, Mwanret D, Queen S (2020) Assessment of the conditions of educational facilities in selected local government areas in Kaduna state, Nigeria. Zbornik Radova Departmana Za Geografiju, Turizam Hotelijerstvo 49–1:81–98

Khan MU (1982) Interruption of shigellosis by hand washing. Trans R Soc Trop Med Hyg 76(2):164–168

Kitchin R, Law R (2001) The socio-spatial construction of (in) accessible public toilets. Urban Stud 38(2):287–298

Kiwanuka M (2002) Response to research questionnaire. Occupational therapist, Community Based Rehabilitation Alliance (COMBRA), Uganda

Kolo MA, Mshelia M, Chutiyami M (2017) Assessment of toilet facilities in selected secondary schools within Maiduguri Metropolitan, North Eastern Nigeria. J Nat Sci Res 7(10)

Kolsky P (2006) Opening remarks for the "Public and community toilets: lessons and ideas from experience in Nairobi, Delhi, Mumbai, and Liuzhou". Video conference, The World Bank

Koopman JS (1978) Diarrhea and school toilet hygiene in Cali, Colombia. Am J Epidemiol 107(5):412–420

Kotloff KL, Winickoff JP, Ivanoff B, Clemens JD, Swerdlow DL, Sansonetti PJ et al (1999) Global burden of Shigella infections: implications for vaccine development and implementation of control strategies. Bull World Health Organ 77(8):651

Lee MJ, Tham KW (2021) Public toilets with insufficient ventilation present high cross infection risk. Sci Rep 11(1):20623

Lidonde R (2005) Scaling up school sanitation and hygiene education symposium. In: The way forward: construction is not enough

Mulholland SJ, Packer TL, Laschinger SJ, Olney SJ, Panchal V (1998) The mobility needs of women with physical disabilities in India: a functional perspective. Disabil Rehabil 20(5):168–178

National University Commission (2006) Self-study form: accreditation of degree programme in Nigerian universities. NUC/SSF, p 18

Officer A, Posarac A (2011) World report on disability. World Health Organization, Geneva, Switzerland

Oyinloye MA, Oluwadare OI (2015) Public conveniences and sanitation challenges in developing nations: a focus on Agege, Lagos, Nigeria. Int J Res Soc Sci 5(7):7–20

Pathak B (1995) History of toilets, paper presented at the international symposium on public toilets held in Hong Kong on May 25–27, 1995. Sulabh International Museum of Toilets

Pierce LL (1998) Barriers to access: frustrations of people who use a wheelchair for full-time mobility. Rehabil Nurs J 23(3):120–125

Restroom Association (Singapore) (2013) A guide to better public toilet design and maintenance, 3rd edn

Sachelo C (2002) Questionnaire response. Executive Director, Zambia National Association of the Physically Handicapped, Ndola, Zambia

Sapey B, Stewart J, Donaldson G (2005) Increases in wheelchair use and perceptions of disablement. Disabil Soc 20(5):489–505

Save the Children (2009) A situation analysis of school health and nutrition in Nasirnagar, Brahmanbaria

Serlin D (2017) Pissing without pity: disability, gender and the public toilet. In: Disability, space, architecture: a reader. Routledge, pp 213–308

Tiswin TN, Luguterah A, Aladago AD (2019) Assessing the types, condition and functionality of water, sanitation and hygiene facilities in public primary schools in the Zabzugu District of Ghana. UDS Int J Dev 6(1):92–101

UNICEF (2007) A report card on water and sanitation: progress for children

United Nations (2010) Including the rights of persons with disabilities in united nations programming at country level

Wada OZ, Oloruntoba EO, Adejumo M, Aluko OO (2020) Classification of sanitation services and students' sanitation practices among schools in Lagos, Nigeria. Environ Nat Resour Res 10(3):55–68

White C, Kolble R, Carlson R, Lipson N, Dolan M, Ali Y, Cline M (2003) The effect of hand hygiene on illness rate among students in university residence halls. Am J Infect Control 31(6):364–370

Chapter 6
Corporate Real Estate Management and the Performance of Listed Business Organisations in Nigeria

Timothy Tunde Oladokun and Bioye Tajudeen Aluko

Abstract The import of the real estate assets of business organisations, otherwise known as corporate real estate (CRE), has shifted from being regarded as a mere factor of production to a strategic business resource which can be used to improve the profitability of the organisation. This study evaluates the contributions of corporate real estate management (CREM) to the performance of listed business organisations in Nigeria. Primary data were collected with the aid of questionnaires served on CRE executives of the 105 listed organisations in Nigeria. In addition, secondary data in respect of the profit margin, share price, dividend payout as well as CREM strategies were also extracted from SEC factbook and the annual reports of the organisations. Regression analysis was carried out to determine the level of CREM contribution to the organisations' profit margin. The result shows that CREM contributed 15.6% and 25.7% to the profit of the organisations in 2010 and 2017 respectively. Omitting the perception of various business organisations could subject the findings of the study to bias. Further research targeted at the customers of the organisations will provide a balanced view. The findings will provide the corporate managers and policy makers with data for efficient utilisation of CRE to business performance and ultimately enhance productivity.

Keywords Business organisations · Corporate real estate · Corporate real estate management · Corporate executives

6.1 Introduction

The need for efficient utilization of organizational resources has severally been established in the literature (Nourse and Roulac 1993; Sarasoja et al. 2004; Lindholm and Gibler 2005; Scheffer et al. 2006). This is usually because, if the main objectives of shareholders, being the achievement of productivity and profitability would be

T. T. Oladokun (✉) · B. T. Aluko
Department of Estate Management, Obafemi Awolowo University, Ile Ife, Osun State, Nigeria
e-mail: toladokun@oauife.edu.ng

S. P. Akinbogun et al. (eds.), *Facility Management practices*,
https://doi.org/10.1007/978-3-031-65321-6_6

achieved, not only must managers be efficient in the use of the available resources, such resources must be consciously used to enhance business lots.

The recent globalisation of business operations, coupled with information technology explosion, requirements for financial transparency as well as competitive strategies are requisite forces influencing the need for efficient utilisation of resources. As such, CRE as an organization's resource is the focus of many studies (Brown 2001; Jan van Ree 2002; Lindholm 2008). Appel-Meulenbroek and Feijts (2007) submitted that CRE, if properly managed, is able to influence performance and also explains the majority of performance changes in business organisations. Effective use of CRE, according to Krumm and de Vries (1998) may provide tangible and intangible impacts on the company which may include effectiveness and efficiency of the activities of the organisation. In addition to the financial benefits of CRE, buildings may advertise and attract attention to a firm's goods and services as well as reflect the purpose of a business and encourage important work relationship.

Whereas studies in the developed world have established that efficient management of the real estate assets of business organisations increases organisations' equity as well as profitability (McDonagh 2002), the contribution of CRE to business organisations in African countries like Nigeria appear not to have been ascertained. This is inspite of the fact that African business market seems to be the focus of foreign investors.

With a shift from perceiving CRE as a cost to the organisation, to a major business resource, it is often now used as an asset that can provide benefits of intangible assets (Lindholm and Gilber 2006). As such, business managers in African countries like Nigeria will be placed on a better pedestal knowing the contribution of CRE to the performance of their overall productivity. Also, with increasing global recognition of the contribution that effective management of CRE can have on the 'overall health' and performance of business organisations, foreign investors interested in Nigerian market will likely hinge their decisions on understanding, to what extent CRE, which is said to be the second largest resource of an organisation, is efficiently utilised to enhance the shareholders objective. Therefore, the study will evaluate the contribution of CREM to the performance of listed business organisations in a developing country like Nigeria. With relatively few studies on the impact of CREM to business performance, especially in African countries, this study will provide information on how real estate can be used to enhance business growth and sustain productivity. The remaining part of this paper has been structured as follows. Following the introduction is the study area and research design. The third section contains the review of past studies in respect of contribution of CREM to business performance. While sections four and five contain the results and the discussions of findings the sixth section contains the conclusions.

6.2 Review of Past Studies

Several studies (see, for example, Nourse and Roulac 1993; Brown 2001; Jan van Ree 2002; Sarasoja et al. 2004; Lindholm and Gibler 2005; Scheffer et al. 2006) have investigated different categories of added value of real estate through the years but with little or no focus of its impact on overall performance. Varcoe (2002) assessed the adoption of performance measurement processes at the portfolio management level by reviewing the practices adopted by USA based group of leading CRE teams. This study, however, did not assess the contribution of CREM to the performance of business organisations. A New Zealand's study of McDonagh (2002) analysed data obtained from a field mail survey of 457 CRE executives from a wide range of commercial and non-profit organisations to evaluate a number of previously identified input factors of performance. Aimed at determining the level of corporate real estate's contribution to overall business performance, the study did not consider the utilisation of appropriate strategies for CRE to contribute to the performance of business organisations.

Liow and Ooi (2004) examined the influence of CRE on shareholders' value using two value-based measures: economic value added (EVA) and market value added (MVA). In evaluating the impact of the high intensity of real estate on the overall value of the organisations, it made no conscious effort to pin point the impact of CREM on the overall performance of business organisations. Lindholm and Gibler (2005)'s work with focus on 26 firms with a range of core businesses in Finland, the Netherlands, the UK and the US, was restricted to the development of key indicators that could be used to measure the performance of CRE.

Deng and Gyourko (2000) and Seiler et al. (2001) examined the relationship between CRE ownership and firm performance for the USA. They documented a negative relationship between real estate ownership and the firm's *b*, but found no significant relationship with firm performance. Louko (2004) investigated the effects of CRE disposals on corporate performance ratios in Europe between the years 1998 and 2002, particularly for the retail and telecom sectors in which some corporations conducted large CRE asset sale-and-leaseback operations. The focus of the studies was limited to the owned portion of the real estate assets, thus making the coverage of the study narrower and the findings non representative of the total real estate assets of the organisations (Appel-Meulenbroek and Feijts 2007).

Brounen and Eichholtz (2005) explored how CRE ownership affects the stock performance of non-real estate companies internationally. They found a negative (but not statistically significant) relationship between relative real estate holdings and risk-adjusted stock performance, with a sector-by-sector analysis showing that the effect of CRE ownership on performance is to a large extent driven by the sector the company operates in. Brounen et al. (2005) showed that, in general, CRE ownership is associated with a strong relative performance for retail companies, in marked contrast to the negative performance effects found for other industrial sectors. The study did not consider the leased portion of the real estate assets of organisations which this study will consider.

In 2006, Lindholm and Nenonen documented a structured approach to developing a real estate strategy in conjunction with the core business strategy supported by a performance measurement system which will allow CRE executives to better communicate how CRE is adding value to the firm, did not reflect the peculiar characteristics of developing countries and left out how CRE influences the performance of the organisations. In Netherland, de Vries et al. (2008) explored the added value of real estate interventions to organisation's performance, theoretically and empirically, including unforeseen positive and negative side effects. The findings noted that the effects are not always as positive as expected, nor are they always clearly visible when using the present KPIs for organisation's performance but however ignored its relevance to organisations in a developing country like Nigeria.

An investigation of the impact of CRE ownership on value creation for non-financial French listed companies using a pool sample of 5BF 250 companies over the period 1999 to 2004 was the focus of Nappi Choulet et al. (2009). While the association between economic added value (EVA) and market added value (MAV) were identified as problems for the value generated by French listed companies and the proportion of real estate in the asset portfolio, the findings which indicated that an increase in the proportion of real estate assets (over total assets) is negatively associated with EVA, did not consider the use of CRE to enhance organisational performance. The study of Jordan et al. (2009) demonstrated how choosing the right business metrics and monitoring them through effective scorecards can identify and amplify the competitive value that CRE creates for its host company.

In Germany, Stadlhofer (2010)'s study with focus on pharmaceutical companies was limited to identifying and quantifying the impact of CRE's performance on shareholders' value and its contribution to core business competitiveness. A recent study of Ristaniemi and Lindholm (2011) evaluated the added value elements of CREM in industrial companies in Finland. The study is faulted on the grounds of failure to consider the aspects of the contribution of CREM to business organisations.

In Africa, the few past pioneering works of Oladokun et al. (2009), Njungbwen and Udo (2010), Oladokun (2010) and Oladokun (2011) have only been limited to exploring the relevance and usefulness of CREM to business organisations in developing countries like Nigeria. The contribution CRE could make to business organisations has largely remained unstudied. In Sweden, the study of Holfert and Villamide (2011) was limited to an examination of the practices of Corporate Real Estate Management with focus on organisations listed in the Nasdaq OMX Nordic, the Stockholm Stock Exchange.

In Melbourne, the work of Bocian and Fortune (2010) investigated corporate real estate outsourcing in the financial sector in Poland from the client's perspective in Australia. Matsham and Heywood (2012) and Fayomi et al. (2019) examined outsourcing practices of corporate real estate management outsourcing. The studies which identified factors such as cost savings and access to skills, technology and best practice as factors that contributed to CREM outsourcing did not look at the impact of it on the performance of organisations. Other studies such as Oladokun (2011, 2013, 2018), Lalloo (2013), Adetokunboh et al. (2013), were all attempts to explore and document the relevance of development of CREM in the African setting.

The foregoing revealed the absence of local studies in respect of the contribution of CREM to business performance in Nigeria as available documented works were carried out in the developed countries such as the UK, USA, Germany, Finland and Singapore with different economic and environmental factors from African countries like Nigeria. Another limitation is that while the contribution of CREM to business performance has been empirically established in those developed countries, its adoption in Nigeria still remains largely a speculation. This study will therefore fill the gap that exists in this regard.

6.3 Research Design/Method

A survey questionnaire method structured to obtain background information about the real estate portfolios of the sampled organisations and the strategies being adopted for the management of their CRE assets was adopted. The study administered questionnaires on senior real estate officers of the one hundred and five (105) organisations that are listed with the Nigerian Stock Exchange (NSE). The focus on senior management within the targeted companies was to ensure not only the accessing of information at an appropriate level but also its accuracy and currency in relation to the contribution of CRE to business performance.

The questionnaire comprised three sections. Section One asked questions about the respondents and their organisations. The questions included job classification, core business, staff strength, geographical spread, educational qualifications, year of experience and professional qualifications. The data was used to depict the appropriateness of the respondents for the study.

Section Two concentrated on obtaining data about the real estate portfolio of the organisations. Questions included number of buildings in portfolio, value of the owned portion of the portfolio, value of the leased portion of the portfolio and annual rental being paid on leased portions of the portfolio. The last section had questions about the views and opinions on the CRE strategies being used. Respondents were also asked to rank on a five point Likert scale their perception about factors influencing the manner by which CRE is utilised in a business organisation.

The one hundred and five (105) questionnaires were administered at the head offices of these organisations. The CRE executives were purposively selected in order to be able to obtain relevant data about the contribution of CREM to the performance of their respective organisations. Out of the one hundred (105) questionnaire administered, a total of sixty eight (68) were returned and found useful. This represents 65% response rate. The data collected were analysed with the use of frequency counts, mean and regression analysis. In addition, secondary data was obtained in respect of the performance of the sampled organisations with the ultimate aim of examining the relationship between the profit margin as well as the real estate management strategies adopted by the organisations. Secondary data in respect of the profit-after-tax, share price, dividend payout ratio, earning per share and total sales per share were extracted from the annual reports of the selected business organisations as well as Stock Exchange Commission (SEC) factbooks for the period of between 2010 and 2017. The idea was to target only organisations that survived the 2005 recapitalisation exercise of the government.

With the profit margin as the dependent variable, the CRE strategies obtained from the questionnaire survey and other financial variables (share price, dividend payout ratio, earning per share and total sales per share) extracted from the annual report and SEC factbook were the independent variable. Linear regression analysis was further carried out to show the effects of the adopted CRE on the profit margin.

6.3.1 The Study Area

With the recent rebasing of the economy, Nigeria can be seen as one of the largest economy in Africa and 26th in the world having/housing both locally owned and multi-national organisations. Notwithstanding the fact that most multi-national companies and large business organisations have national geographical spread, the headquarters of most of those organisations are in Lagos State (Babawale and Nubi 2011). This is not unexpected as Lagos is the Nigeria's economic and financial capital. As Africa's second fastest growing urban centre after Cairo and the focal point for regional, national and international commercial activities. It might also not be out of place to relate the citing of the corporate headquarters of organisations in Lagos State as a means to benefit from the economic advantage of both the air and sea ports which serve to link the country with global economic world.

6.4 Data Presentation, Analysis and Discussion

In presenting the results of the questionnaire survey, the profile of the respondents is presented first, followed by analysis of the impact of CREM on the performance of business organisations in Nigeria. Both primary and secondary data were required to accomplish the study. Questionnaire was designed to obtain records on CREM strategies adopted as well as to evaluate the contribution of CREM to the overall performance of the organisations.

6.4.1 Company Profile

The study examined the profile of the sampled organizations in line with the global changing business structure/terrain so as to establish their preparedness for strategic real estate management. The result of the analysis is discussed in the following sections Adopting the classifications of listed firms used by the Nigerian Stock

Exchange, the analysis revealed that the highest response rate (22.05%) came from the insurance companies, followed by the banking industry (17.12%). Organisations in the food/beverages sector had (16. 18%) while the manufacturing and the chemical and paints sectors each had a response rate of 4.41%. The least response rate (2.94%) came from the construction sector. The result revealed a good representativeness of the listed organisations in the country which the study can rely upon for generalisation.

The average number of people employed by the organisations was over 500 with majority of the firms operating from more than 100 branches and locations in the country. The organisations can therefore be regarded as big with full national coverage.

Further findings showed that majority (84.7%) of the CRE executives had Bachelor of Science (B.Sc.) as their minimum qualifications with an average length of twelve (12) years of experience. The responses showed that 61% of the CRE executives were Associate Members of the Nigerian Institution of Estate Surveyors and Valuers (NIESV). A sizeable (30%) were probationers of NIESV while 8% represented the fellows of the NIESV.

The above results showed that most of the respondents were adequately trained and their level of experience suggested that the organisations employed skilled labour who added value to their businesses.

Findings from Table 6.1 showed that the sampled organisations had substantial real estate holdings. Sixty five (65%) of the organisations had more than 100 properties while 28.0% had less than 50 properties. The remaining 7% had between 51 and 100 properties. Their real estate portfolio was classified into residential, commercial (offices) and warehouse categories.

The result on the ownership status of those properties showed that the organisations owned more than 70% of their property portfolio while the remaining 30% were taken on lease apparently to avoid business disruptions that characterise lease structuring and administrations especially when leases are not renewed at expiration. It could also be justified by the desire to use real estate to boost the value of their organisations on the balance sheet.

The result on the value of their owned properties revealed that 57% of the sampled organisations owned property assets that is worth more than N600 million. While the majority (17%) of the organisations paid over N300 million on the leased portion of their CRE, the annual rental paid by the remaining organisations ranged between N50 million and N300 million per annum. The findings suggest that the CRE of Nigerian business organisation is substantial enough to warrant or attract management attention.

Table 6.1 Sampled organisations' real estate portfolio profile

	Real estate profile	Percentage (%)
No of buildings in portfolio	< 50	28.0
	51–100	7.0
	Over 100	65.0
Total		**100.00**
Value of owned portion of portfolio	Less than 100 million	10
	100 m–200 million	6
	201 m–300 million	6
	301–400 million	2
	501 m–600 million	2
	Over 600 million	57
Total		**100.00**
Value of leased portion of portfolio	Less than 50 million	16
	51 m–100 million	22
	101 m–150 million	4
	151–200 million	8.2
	201 m–250 million	4.0
	251 m–300 million	8.0
	Over 300 million	17.0
Total		**100.00**
Owned portion of portfolio	Average	73.0

6.5 Contribution of CREM to the Financial Performance of Business Organisations

The study evaluated the nature and the degree of the contribution of CREM to the performance of business organisations.

In achieving this, three sets of analysis were carried out in line with the study of Ali et al. (2008) in order to be able to demonstrate the effect of CREM on the performance of business organisations. First, linear regression analysis was carried out with year 2010 financial data (profit margin) as the dependent variable and the CREM strategies as well as other financial variables of share price, earning per share, volume of sales per share price as control variables. A similar analysis was carried out for year 2017. The third analysis regressed the profit margin of 2017 against the CRE strategies adopted in 2010 so as to show the effects of time lag on the contribution of CREM. The result of the analysis of data in respect of the contribution of CREM to the financial performance of the business organizations are presented in the following sections.

Table 6.2 Linear multiple regression showing the relationship between 2010 profit margin and 2010 CRE strategies

Model	R	R^2	Adjusted R^2	Std. error of the estimate
1	0.395[a]	0.156	0.004	1.72546E5

[a]*Predictors* (constant), support public relation and marketing, share price, increase of value, sale per price, change the culture, support public relation and marketing, earning price, reduce cost, increase flexibility
[b]*Dependent variable* profit after tax (Profit after tax represented the profit margin upon which the financial data were regresses)

6.5.1 Linear Regression Showing the Relationship Between 2010 Profit Margin and 2010 CRE Strategies

As a first step, the profit margins recorded by the organisations in 2010 were regressed against the CRE strategies they adopted in the same year. The result of the analysis is contained in Table 6.2.

The result, as contained in Table 6.2, reveals that the R^2 value of the regression analysis was 0.156. If expressed as a percentage, the result indicated that the adopted CRE strategies accounted for 15.6% of the achieved profit of the organisations. The result showed a slight and minor impact of real estate on the financial performance of the organisations. The result indicated that CREM contributed about 15.6% to the profit of the organisations in 2010.

6.5.2 Model Summary Showing the Relationship Between 2017 Profit Margin and 2017 CRE Strategies

The study further evaluated how much of the variance in the dependent variable (profit after tax) was explained by CREM in 2017. The result is contained in Table 6.3.

The result, as contained in Table 6.3, revealed that the R^2 value of the regression analysis was 0.257. The indication is that CREM strategies adopted in 2017 explained 25.7% of the variance in the profit of the organizations in 2017. This result showed a relatively higher and more impressive contribution of real estate to the financial

Table 6.3 Linear regression showing the relationship between 2017 profit margin and 2017 CREM strategies

Model	R	R^2	Adjusted R^2	Std. error of the estimate
2	0.507[a]	0.257	0.076	12,876.91655

[a]*Predictors* (constant), capture real estate value, share price, sale per share, facilitate and control production, earning per share, promote sales and selling process, flexibility, promote human resource objectives, facilitate managerial process, occupancy cost minimization, promote marketing strategy
[b]Dependent variable

Table 6.4 Summary of linear regression showing the relationship between 2017 profit margin and 2010 CRE strategies

Model	R	R^2	Adjusted R^2	Std. error of the estimate
3	0.425[a]	0.180	0.101	12,605.50493

[a]*Predictors* (constant), promote marketing strategy, sale per share, earning per share, share price, flexibility
[b]*Dependent variable*

performance of the organization. The relative contribution of more than 25% by CREM strategies to the profit of the organisations could explain a more adoption and mastery of CREM strategies by the organisations after seven years of full adoption.

6.5.3 Model Summary of Linear Regression Showing the Relationship Between 2017 Profit Margin and 2010 CRE Strategies

Further analysis was carried out in order to allow for the effects of time lag between adopted CRE strategies and financial performance of the organisations. The aim was to explore the potential for a lagged relationship between the 2010 corporate strategies employed by the companies and financial performance as measured by profit margin in 2017.

Thus, the study evaluated how much of the variance in the dependent variable (profit after tax) is explained by the CRE strategies in 2010, should the organisations adopt the same strategy. The result is contained in Table 6.4.

As contained in Table 6.4, the result is that the R^2 value of the regression analysis was 0.180. The result in Table 6.4 showed that CREM explained 18.0% of the variance in the profit of the organizations thus showing a slight and minor impact of real estate on the financial performance of the organizations if the organisations have not changed their CRE strategies. It could then be said that organisations who did not change the CREM strategies adopted in 2010 would only record a contribution of 18.0% to the profit as against 25% contribution by the currently adopted strategies in 2017.

6.6 Conclusion

The study evaluated the contribution of CREM to the financial performance of listed business organisations in Nigeria. The study established that Nigerian listed business organisations have substantial real estate holdings and are being largely managed by competent and trained professionals. It was equally established that a marginal proportion of the financial performance of the organisations is explained by the use and adoption of real estate strategies to drive the organisations.

A major conclusion from the study is that Nigerian listed business organisations are benefitting from efficient real estate management. Efficient CREM has proved to improve the lots of the organizations and place them on a better pedestal for improved productivity.

A policy implication from the study is the need for business executives to integrate real estate into the main stream of business activities. This can be achieved by ensuring that real estate study is made an integral part of business executive programmes. Understanding the roles RE can place in business environment will make business executives place more priority to the management of the asset for competitive advantage.

References

Adetokunboh O, Aibinu I, Agbato S (2013) Corporate real estate development in Nigeria. Issues and challenges of site acquisition. Paper presented at the Africa real estate society conference in Kigali, Rwanda from 23–26 October 2013

Ali Z, Mcgreal S, Adair A, Webb RJ, Roulac ES (2008) Corporate real estate strategies and financial performance of companies. J Prop Res 25(3):241–267

Appel-Meulenbroek R, Feijts B (2007) Corporate real estate effects on organizational performance: measurement tools for management. J Corp Real Estate 9(4):218–238

Babawale GK, Nubi TO (2011) Property tax reform: an evaluation of Lagos State land use charge, 2001. Int J Law Manag 53(2):129–148

Bocian R, Fortune J (2010) Corporate real estate outsourcing in the financial sector in Poland: the client's perspective. Available from Sheffield Hallam University Research Archive (SHURA), pp 70–86. http://shura.shu.ac.uk/7581/

Brounen D, Eichholtz P (2005) Corporate real estate ownership implications: international performance evidence. J Real Estate Finan Econ 30(4):429–445

Brown MG (2001) Choosing a company's building design: models for strategic decisions. J Real Estate Res, vol 22 Nos 1/2, pp 81–105

Brounen D, Colliander G, Eichholtz MA (2005) Corporate real estate and stock performance in the international retail sector. J Corp Real Estate 7(4):287–299

Deng Y, Gyourko J (2000) Real estate ownership by non-real estate firms: an estimate of the impact on firms returns, working papers 321. Wharton School Samuel Zell and Robert Lurie Real Estate Center, University of Pennsylvania, Philadelphia, PA

De Vries JC, de Jonge H, van der Voordt JM (2008) Impact of real estate interventions on organisational performance. J Corp Real Estate 10(3):208–223

Fayomi I, Oladokun TT, Adewolu TO (2019) Factors influencing corporate real estate outsourcing in Nigeria: an empirical study from the banking and telecommunication sectors in Lagos, Nigeria. In: Proceedings of 1st international conference of engineering and environmental sciences (ICEES), Osun State University, pp 257–272, holding from 5th–7th November

Jan van Ree H (2002) The added value of office accommodation to organisational performance. Work Study 51(7):357–363

Jordan M, McCarty T, Velo B (2009) Performance measurement in corporate real estate. J Corp Real Estate 11(2):106–114

Holfert E, Villamide M (2011) Corporate real estate management practices in Sweden, master thesis submitted to the department of real estate and construction management, royal institute of technology, Sweden

Krumm PJMM, de Vries J (1998) Value creation through the management of corporate real estate. J Property Investment Finan 21(1):61–72

Lalloo A (2013) Corporate real estate practices in South Africa—a survey of the top 200 companies listed on the Johannesburg stock exchange. A dissertation report submitted to the Faculty of Engineering and the Built Environment, University of the Witwatersrand, in fulfilment of the requirements for the degree of Master of Science in Property Development & Management

Lindholm A, Gibler KM (2005) Measuring the added value of corporate real estate management—beyond cost minimization. A paper presented at the 12th annual European Real Estate Society conference, Dublin, Ireland, June 15–18 2005

Lindholm AL, Gibler KM, Levaïnen KI (2006) Modeling the value-adding attributes of real estate to the wealth maximization of the firm. J Real Estate Res 28(4):445–474

Lindholm AL (2008) Identifying and measuring the success of corporate real estate management. Doctoral dissertation, Helsinki University of Technology, Finland

Liow KH, Ooi JTL (2004) Does corporate real estate create wealth for shareholders? J Property Investment Finan 22(5):386–400

Louko A (2004) Corporate real estate disposal impact on corporate performance ratios. Int J Strateg Prop Manag 8(3):131–147

Matsham A, Heywood C (2012) An investigation of corporate real estate management outsourcing in Melbourne. In: Eighteen Pacific real estate society conference, Adelaide, Australia, pp 15–18

McDonagh J (2002) Measuring corporate real estate management performance. In: Seventh PRRES conference, Canterbury, pp 21–23

Nappi-Choulet I, Missonier-Piera F, Cancel M (2009) Value creation and the impact of corporate real estate assets, an empirical investigation with French listed companies. J Corp Real Estate 11(2):79–91

Njungbwen E, Udo GO (2010) Benefits of corporate real estate management to higher education institutions. 1st African international business and management [AIBUMA] conference on, knowledge and innovation leadership for competitiveness. Kenyatta International Conference Centre, Nairobi, Kenya 25th and 26th August Available [Online] available at: www.aibuma.org/proceedings/downloads/Udo,%20Emem,%20Nigeria.doc. Accessed 03/01/17

Nourse H, Roulac E (1993) Linking real estate decision to corporate strategy. J Real Estate Res 8(4):475–494

Oladokun TT (2010) Toward value-creating corporate real estate assets management in emerging economies. J Property Investment Finan. Emerald Group Publishing, UK. 28 (5): 354-364. (Nigeria). https://www.emerald.com/insight/content/, https://doi.org/10.1108/14635781011069963/full/html

Oladokun TT (2011) Corporate site selection and acquisition in a Nigerian GSM communication company. J Corp Real Estate 13(4):247–260

Oladokun TT (2013) Corporate real estate management practice in Nigeria: a survey of the attitude of business executives. Ife J Sci Technol 1(1):27–32

Oladokun TT (2018) Challenges of corporate real estate outsourcing in Nigeria. J Int Real Estate Constr Stud 4(2):111–121

Oladokun TT, Aluko BT, Odebode AA (2009) Corporate real estate management: a need for paradigm shift in Nigeria. The Estate Surveyor Valuer, J Niger Inst Estate Surveyors Valuers 32(1):67–73

Ristaniemi E, Lindholm A (2011) Added value of corporate real estate management in industrial premises [Online] at: www.cres.scix.net/data/works/att/eres2011_134.content.pdf. Accessed 03/01/17

Sarasoja A, Gibler KM, Levainen Kl (2004) Value adding attributes of CREM, Andrew young school of policy studies, urban and regional analysis group (URAG) working paper no. 04–09, [Online] Available at: http://aysps.gsu.edu/urag/workingpapers/index.htm. Accessed 03/01/17

Scheffer JJL, Singer BP, Van Meerwijk MCC (2006) Enhancing the contribution of corporate real estate to corporate strategy. J Corp Real Estate 8(4):88–197

Seiler M, Chatrath A, Webb J (2001) Real estate ownership and the risk-return to stockholders. J Real Estate Res vol 22 Nos 1/2, pp 199–212

Stadlhofer G (2010) Corporate real estate performance: contribution to core business competitiveness at global pharmaceutical enterprises. J Corp Real Estate 12(2):96–116

Varcoe B (2002) The performance measurement of corporate real estate portfolio management. J Facil Manag 1(2):117–130

Chapter 7
Assessing the Viability of Care Home Facilities in Lagos, Nigeria

Oluwafunbi K. Medale and Nnaemeka B. Ezeokoli

Abstract Senior or care homes for the elderly are a relatively new phenomenon in Nigeria, and they have not received much attention as a real estate investment option. This paper aims to examine the viability of investing in senior or care homes for the elderly in Lagos State, Nigeria. It examined the viability of the investment and the factors affecting the residence of the elderly as an investment in the area of study. Data gotten through structured questionnaire administered on four (4) investors operating facilities for the elderly in Lagos, Nigeria was analyzed using descriptive analytics (Weighted Average Scores, frequency distributions, and a graphs) and discounting viability appraisal techniques [NPV, IRR, and Profitability Index (PI)]. Factors such as affordability, Governmental and cultural factors ranked high among the major factors affecting old peoples' home investment in the area of study. The findings further revealed that old peoples' home is a novel, yet viable investment that require huge capital outlay, which is the reason small scale investors could not easily break into the market. As a result, government should formulate legislations that favour the growth of this industry. In addition, Prospective investors are encouraged to take advantage of this investment among other various real estate investment options.

Keywords Appraisal · Facility · Investment · Old people's home · Viability

7.1 Introduction

Ageing comes with its own physical, mental, psychological, and health wise challenges, and it is an inevitable phase of life (Ugwu and Ugwu 2007). Any action(s) taken to counter this occurrence is key to improving the healthy lifestyle of an older person. Statistics show that those aged 65 years and above (the elderly) in Nigeria make up 3.1% or 5.9 million of the total population, which in rough account represents an increase of 600,000 between the year 2012 and 2017 (Population Reference Bureau 2012). Thus, various nations around the world especially the western world

O. K. Medale · N. B. Ezeokoli (✉)
Department of Estate Management, Federal University of Technology, Akure, Nigeria
e-mail: nbezeokoli@futa.edu.ng

S. P. Akinbogun et al. (eds.), *Facility Management practices*,
https://doi.org/10.1007/978-3-031-65321-6_7

consider investment in senior/care or old peoples' home facilities an important sector of their economy where older persons are cared for, while their families go on with their normal lives.

The Old peoples' home is an institutional facility designed to cater for the old in age, usually from 60 years and above (Oyinlola and Oyinlola 2017). It is a retirement and old people nursing home with well-structured property built to meet physical, mental and medical needs of the older generation. According to Aboderin (2016), a standard old people's home should meet all the facilities for meals, gathering, recreational facilities, transport and medical services. The old people's home is traditionally not an African concept rather, a western culture; Nigerian way of managing older persons is rooted in the extended family (Okoye 2013). This has brought advent of private and public organizations investors to invest into the sector. The World Health organization policy framework on active ageing portray maximum care for the old as: maintenance of state of good health, participation in social activities, and all round security, which promote optimal health and well-being during the aging process. The provision of facilities like old people's home that can shoulder all these aforementioned responsibilities, of maximum attention and care needed by older persons in the country is an area of great concern. However, in this part of the world, not so much work on the old people's home facility and its profitability as an investment has been done.

Scholars around the world have written on different areas of the old people's home concept and management. Most of these studies emphasized old peoples' rights, housing for old people and believe system of old peoples home especially in Africa. There is minimal research done on old people's home as an investment opportunity especially in Nigeria. Oyinlola and Oyinlola (2017) did a study on misconception and conception of old people's home in Nigeria i.e. the believe of Nigerians on putting old people in a facility rather than taking care of them at home by family and relatives. The study found out that old people generally live and receive care in their homes or the residence of their children and relatives. This however is not saying that the care is satisfying or that there are no cases of neglect of the aged.

The believe system of Nigerians, the wrong idea of old peoples home as a destitute abode, and the willingness of the old to live in places order than their homes are factors that can inhibit an investor from investing in old peoples home as an investment (Akinny 2016). Nevertheless, there are no empirical evidence to ascertain this fact especially in a country that does not have a working legislation for the housing and welfares of its old. Osunderu and Adeyoyin (2018) examined the elderly abuse and care of the aged in Nigeria society, a case of old people's home facility in Yaba using three fundamental variables for comparison. The percentage of persons who do not see any significant reason in keeping aged people in old people's home was 59.5%, while the percentage of population who thought it fit to do so was 40.05%. However, the study was limited to why general populace do take their aged to the old people's home and the implication of government policies on the care of aged, whereas this study seeks to examine how these human and governmental factors influence the old peoples' home viability as an investment in Nigeria.

According to United Nations (2006), the population of the older generation in Nigeria is on a sporadic increase with the statistics of 26% increase as at 2020 and predicted 30% increase by the year 2050. This makes old peoples' housing and welfare a concept that should not be ignored by either the government or the real estate property investors. However, Aboderin (2004) revealed that the care that the older ones enjoyed from the family members and relatives has decreased over time in Nigeria. This decline was because of a shortage in financial capacity of the young adults. It further revealed that there is decline in family support to the older persons, thus, it posited that this generation has swept the principle of care reciprocity in Nigeria culture under the grass as older persons now depend on themselves. Few scholars in this area of study in the western world support the need for older persons to live in an old people's home facility rather than relative homes. For instance, Wang and Shi (2008) stated that bereaved older persons are healthier in nursing facility compared to those who live in isolation. In other words, older persons who live in isolation are more in need for emotional and daily personal care, and since old people's home can provide the varieties of care (personal, medical and psychological); it appears to be the best place to spend their retirement (Liu 2010). Hence, the need for old people's home facility to be a suitable option for the older persons in the urban area (Li and Ban 2015). The reality of these factors make the provision of good standard and functional property facilities for the adequate care and attention starved by the older persons a subject of concern.

According to the report of Gov.UK (2017), Old peoples' home is worth 13.5 billion dollars (over Six Trillion naira) with 11,300 Service Care Provider facilities in the UK. Whereas the older person's population in Nigeria, according to the National Population Commission in 2006 was estimated to be over 9 million which may have risen to over 20 million todays, with only 25 service care facilities provider including government, private and NGOs. It is therefore very evident that there is a gap to fill and a demand to meet by the investors. In Nigeria, community-based care was proposed as a policy option. Community based care is geared towards encouraging the elderly to remain at home. According to Okoye (2004), if specialized institutions that caters for the older persons in Nigeria were in existence, many older persons would have been better cared for. Because the concept of old peoples' care in Nigeria is vested in their family members and relatives, it becomes a question of whether there will be demand for old peoples' homes, and how viable such investment will be.

It is therefore pertinent to look into the viability prospect of old peoples' home as an investment for future investors in real estate. Although there are some of these old people's home in existence in Nigeria, but the area of its viability to investors has not been researched into, so also the factors that can inhibit the old peoples home investment in Lagos state, Nigeria. The study examines the factors influencing investment in old people's home in Lagos, evaluate the viability of investing in old people's home facilities in the study area.

7.2 Concept of Old People's Home

Old peoples' home is a residence where old people live and are cared for when old age prevents them from looking after themselves in their own homes. Accordingly, it is a facility built for dual purpose as an institution; a home with health care provisions for a good standard of living and an accommodating environment. It is an extension of a home rather than a health facility. However, the inevitable difference in the world's economic, political and cultural practices have created the diverse approaches to the care of older persons (Akinny 2016), though these differences do not justify the inadequate care given to their welfare, especially in African.

In the sub-region of African, the factors that continually inhibit the poor welfare of the persons are the same across the region. The factors which include poverty, cultural belief, poor pension plan for the older generations and incessant rate of urbanization, which cut across African countries like Nigeria, Malawi and Ghana.

7.2.1 Type of Old People's Home

The old peoples' home can take varied forms as it has diverse meanings to different older persons, which solely dependent on their present physical state (Róin 2015). Old peoples' home is specifically designed to accommodate and deliver an effective care to the aged ones, and also to provide a safe and friendly environment. Due to the peculiarity of the needs of the aged ones, as some need more help than others, especially when there is a life threatening diseases associated with ageing, there exist different types of old people's home according to the services they render. Care homes is a type of old people's home that gives the required personal care to its resident such as washing, medication, supervision, dressing and help to the convenience, and somehow also encourage outdoor activities (Akinny 2016). Care homes have proven to be a good option for families who seek for a more finite solution to the burden that comes with old age. Age UK (2020) identified type of care homes as follows: Care homes with dementia (Care home specifically designed for the purpose of making the old with dementia feel secured and comfortable as the diseases often affect the older persons), Dual-registered care homes (the combination of purpose built facility for personal care and nursing care home), Independent living apartment (this is a type of apartment suitable for older persons who do not require more care personally or medically but would love to be in an environment with their peers. The facility is built to provide just personal care and engage them in excursions, outside trips to the parks, physical fitness, this type of apartment is not licensed or regulated); Adult Homes is a temporary or long term residence for old people for older persons who are unable to live independently. They provide supervision, personal care, house keeping with all round day meal preparation.

Nursing homes on the other hand was described by Nyanguru (1990) in a study of the quality of life of the elderly living in institutions and homes in Zimbabwe, as the type of homes with in-house nurses and workers trained and qualified professionally for sole purpose of caring for the aged. It further explained that their services are strictly personal care to its residents and also assist in medical care. It is otherwise called nursing home for elderly persons who seek for long-term care institutions. The study argued that nursing home is a place where survival is ensured. The facilities in these homes slow down the rate of Ageing, helps to maintain their core strength, and restores their lost functioning (Nyanguru 1990).

Assisted Living home was explained by Age UK (2020) as a home with combined facilities apartment style housing, organized for social interaction, and private duty support services. Health care is often outsourced, this setting is design for old persons who needs assistance with the routine work daily e.g. meals, supervision for medication, physical helping hands, bathing, dressing and transportation. This house also accommodates residents with Alzheimer's and other health challenges pertinent to older persons in the society. Similarly, in a study to explore people's attitudes and preference for assisted living facilities in comparison to nursing homes. The study found out that assisted living facilities were most preferred. It furthered revealed that assisted living facilities is more hospitable than a nursing home (Imamoğlu and Imamoğlu 2006).

7.2.2 *Viability of Investment in Old People's Home*

Reilly and Brown (2006) defined investment as the act of receiving a rate of return which will compensate the investor for the period the capital was invested, adjusted for inflation and the uncertainty involved through an initial commitment of funds. Investment decision of an investor is a compromise between future and immediate consumption. Laing and Buisson (2013) in a study of care of elderly people in the United Kingdom market found that within the year 1990 and 2008, old people's home facilities grew more profitable than the Non-Governmental charity organization sector by 85% margin, which remained constant during the period. Yang (2017) agree with this notion that the older persons' home investment in the western world is a successful endeavor, and that the guarantee of long-term return on such investment solely depends on their high occupancy rate. In Australia, old people's homes investment has proven to be a fast growing viable investment to investors according to annual survey of aged care homes by Bentley chattered accountants (Allard 2016). According to Franchise Business Review (2012), the old people's home industry is one of the five most lucrative sectors in United States of America.

7.2.3 *Factors Influencing Investment in Old Peoples Home*

According to Oladeji (2011) and Odaman and Ibiezugbe (2014) in a study of Health behavior among the elderly in Edo state, Nigeria, the old tend to struggle with identity to fit into the new life, which is an inevitable reality after retirement. It is a war adjustment from the position of being an occupied citizen to a dependent. The bureaucracy in the government administration for accurate and timely pension is another issue that might send this set of people to old age depression, delay in payment of pension or no payment at all, low standard of living and inadequate pension regulation enforcement (Apere 2015). A facility and environment with same peers can perhaps reduce the early pressure from retirement, but most of these retired population are not aware of this possibility or does not have the financial capacity to enroll in one.

Misconception that the residents of old peoples' home is filled with sick and dying, more also that the residents in old people's home often smell bad (Oyinlola and Oyinlola 2017). Similarly, Nyanguru (1990) agreed to that notion that residents in old people's home have been described as disoriented, disorganized, withdrawn, apathetic, depressed and hopeless. Oyinlola and Oyinlola (2017) further explained that the cultural believe system that subject some willing children and relatives of an older person to guilt, if they decide to consider old peoples home as an option for giving their old best care. This is a major factor that can affect old people's home as an investment opportunity in the study area.

Cumming and Henri (1961) argued in a study of Growing old: The process of disengagement in New York City that Social disengagement can be explained as a mutual withdrawal and disengagement of the old from the family and the outer environment. However, the lackadaisical attitude of the government towards providing good welfares, which includes good housing and health care system for the senior citizens and good policies are issues of concern for this area of investment (Ajomale 2007). Hence, Older person's population are not supposed to be undermined by the government, as policies on health care and welfare made by the government should cater for these group of people first just like the lower dependent, but reverse is the case (Oladeji (2011)

Environmental factors such as the disposition of people to older persons in the society. For instance, resentment attitude of a caregiver (Oyinlola and Oyinlola 2017). Furthermore, according to Oladeji (2011), more than 50% of older persons in Nigeria are not secured economically and socially. As the statutory age of retirement in Nigeria is between 60 and 70 years depending on state and institution, after retirement, older persons suffer from the economic productive phase and thus experience economic hardship, since poverty remains a major challenge in Nigeria (Tanyi et al. 2018). Similarly, Odaman and Ibiezugbe (2014) posit that the opportune older persons living in cities of Nigeria only make use of health services when they are mostly affordable and accessible. Hence, majority of older persons who which to live in the old peoples home rather their own cannot afford these luxury unless helped by a willing third party, government or charity organization (Fodder and Allan 2011).

Fukui et al. (2014) examined the determinants of financial performance of home–visit nursing agencies in Japan. The study identified factors that can pose a threat to viability of old people's agencies home care as staff employment, patient utilization, quality control of care, cultural believe and financial capacity. It further explained that quality management culture on the facilities is important to achieve stable financial returns. On the other hand, in the United Kingdom Fodder and Allan (2011) claims that with over 10,000 old people's home in the UK, by most general measures, the organization increases every year, which give rise to high level of competition within the market unlike the Nigeria situation, which is still at its lowest ebb. Hence, the need to carry out this study.

7.3 Research Method

Old people's home investment has been an area of neglect in the Real Estate Industry in Nigeria, hence, the need to target this class of property investment. A survey of old people's homes investment in Lagos shows that there are five (5) privately established old people's homes in Lagos State. This constitutes the sampling frame for the study as well as the sample size as proposed by Israel (2002) for small population less than 200. However, out of these five (5) investment properties, response came from four (4) of the investment properties representing 80% of the sample size. The study relied on primary data source through structured questionnaire designed to answer the objectives of this research. Two categories of Old People's Home investments were considered, which are investments based on lease agreement and those on freehold basis. The lease hold arrangement is a situation where the investor takes up the facility used, especially the building, on lease from a landlord, and pays rent according to the lease agreement. Whereas, the freehold basis has to do with the investor developing the building for the purpose of the business and using the property for same. Rents are not being paid annually on this category of property investment. The analysis of the data collected was carried out using descriptive analysis such as percentage frequency Tables to describe the demographic distribution of the respondents, type of old people's home and different type of services they provide. Weighted mean score was used in analyzing the factors influencing the viability of the care home or old people's home investment in Lagos state, Nigeria. The viability of the investments was analyzed using discounting appraisal techniques such as the Net Present Value, Internal Rate of Return and profitability Index.

Net present value: The net present value of a project express as a sum of money representing the discounting present value of the flow of income, less discounted value of all outgoings, including capital outlay obtained from the investment over the whole of its life. Unlike the payback period, the net present value considers cash flow after payback period, and also take into account the time value of money.

Formula:

$$NPV = \sum_{t=1}^{n} \frac{CF_t}{(1+i)^t} - I_o$$

CF_t cash-flow in time period t
i discount rate or cost of capital
I_o initial capital outlay

n project's expected life

Internal Rate of Return (IRR): Internal rate of return is the rate of return, which equates the present value of the future cash inflows to the present value of e cash outflows. IRR may be ascertained by trial error i.e. discounting the annual receipt at different rate until the rate, which make NPV equals zero is found or graphically.

Formula:

$$NPV = \sum_{t=1}^{n} \frac{CF_t}{(1 + IRR)^t} - I_o = 0$$

Profitability Index: It is a capital budgeting decision criterion defined as the ratio of the present value of the future free cash flow to the initial capital outlay. This method is known as the "Benefit Cost Ratio" in the sense that it is used in evaluating the economic worth of a capital investment projects. The technique evaluates the return on every one Naira invested into a project. It tells how much a project has generated.

Formula:

$$PI = \frac{\sum_{t=1}^{n} \frac{CF_t}{(1+i)^t}}{I_o}$$

CF_t cash-flow (-ve or + ve) in time period t
i discount rate or cost of capital
I_o initial capital outlay
n project's expected life

7.4 Data Analysis and Discussion of Results

This section of the study presents analysis of data collected from the study area and the discussion of results. The analysis was structured to investigate the viability of old people's home as an investment in Lagos state, Nigeria, and the factors affecting this investment opportunity in the study area.

Information on the socio-economic characteristics of the respondents to establish the relevance of the retrieved data to the study is presented in Table 7.1.

Table 7.1 shows the distribution of respondent status according to their year of establishment, official designation, qualification and experience of respondent of the old people's home in Lagos, Nigeria. From the Table, 75% of the establishments have been in existence for more than 10 years, while 25% has establishment year of less than 6 years. The Table further showed that 50% of the respondents were Head of Department, while 25% are Chief Executive Officer and General Manager respectively. Also, 75% of the respondents are Bachelor Degree or Higher Diploma degree holders, while 25% of holds Master's degree. In addition, the Table shows that 75% of the respondents have more than six (6) years' experience in the industry. This

Table 7.1 Socio-economic characteristics of old peoples home investors in Lagos, Nigeria

Socio-economic characteristics	Distribution	Frequency	Percentage (%)
Year of establishment	1–5	1	25
	11–15	2	50
	16–20	1	25
Official designation	Chief executive	1	25
	General Manager	1	25
	H.O.D	2	50
Academic qualification	HND/BSC	3	75
	MSC	1	25
Experience of respondent's	1–5	1	25
	6–10	2	50
	11–15	1	25
Total		**4**	**100**

Source Field Survey (2021)

implies that majority of the investments have been in existent long enough to provide a reliable cash flow for the analysis for at least 10 years. Most of the respondents in the establishments are high ranked officers with minimum academic qualification of Bachelor degree, and have worked with the establishment for more than a five years. This implies that the information and data for this study are reliable and can be used for making inference.

Information on the various services provided in Lagos old people's home, which also can be likened to type of old people's home that exists in the study area are presented in Table 7.2.

Table 7.2 shows that care home was a basic function of old people's home, while other services such as nursing home, adult home, home care with dementia, retirement home and assisted living home were additional functions or services that most of the old peoples' home incorporated into their services. Although, the investment that does not offer these other services, occupies a property on rental basis, which would

Table 7.2 Services rendered by old peoples home in Lagos

Services	Frequency	Percentage (%)
Nursing home	3	75
Adult home	3	75
Care home	4	100
Care with dementia	3	75
Retirement home	3	75
Assisted living	3	75

Source Field Survey (2021)

explain why its service is limited to the basic function of care home because basic care for the older person might not require a purpose built facility for such.

Table 7.3 shows the descriptive statistics of Old People's Home Investment Returns in Lagos and this includes the mean, median among others of both investments on rental basis and freehold basis. From these statistics, it can be observed that the average annual net income of old people's home investment on rental basis is ₦6,854,545, while that of freehold basis is ₦32,042,636. This implies that there is higher net annual income from old people's home investment on freehold basis than that on rental basis. This is reasonable because people prefer to buy lands and build on it to leasing a land for home construction (Kehinde et al. 2021). The Skewness, Kurtosis, and Jarque–Bera values indicate that the distribution of the old people's home investment on both rental basis and freehold basis is not normally distributed; the rental basis skewed left and freehold basis skewed right. Moreover, rental basis data is still closer to normal distribution than that of the freehold basis data. This suggests that the rental basis investments are around the mean value compared to that of the freehold basis investment, thus indicating less variability (risk) than that of freehold basis (Fig. 7.1).

The trends of net income indices for both investments show that the investment on rental basis has a relatively stable income index than the investment on Freehold basis. The stability in income can be due to payment of rent and other charges in the course of running daily activities. On the long run, the investment on freehold basis can also witness relative stability in income. Finally, a sharp rise in income in the year 2020 shows that the Lagos old people's home investment thrive even better in the midst of the pandemic that warranted a lock down and closure of many other investments.

Table 7.3 Descriptive statistics of old peoples' home investment returns in Lagos

	Old peoples' home investments	
	Investment on rental basis (A)	Investment on freehold basis (B)
Mean	6,854,545	32,042,636
Median	7,400,000	30,490,000
Maximum	10,000,000	60,000,000
Minimum	3,800,000	11,108,000
Std. Dev	2,198,801	13,703,206
Skewness	− 0.197883	0.496551
Kurtosis	1.622065	2.822252
Jarque–Bera	0.942028	0.466513
Probability	0.624369	0.791951
Sum	75,400,000	3.52E+08
Sum Sq. Dev	4.83E+13	1.88E+15
Observations	11	11

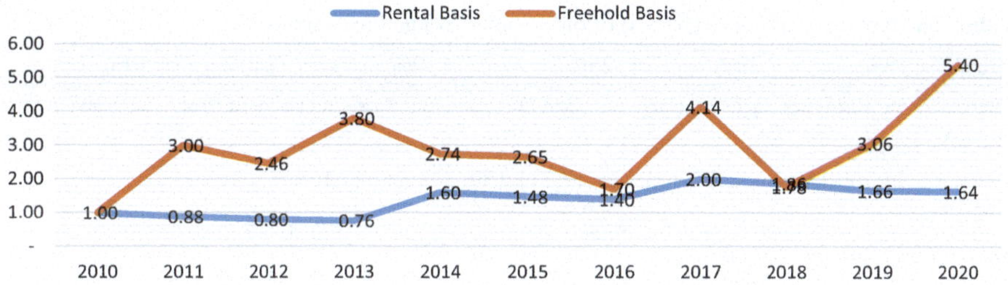

Fig. 7.1 Trends in net income indices of old people's home on rental and freehold basis

From Fig. 7.2, the percentage change in the net income of old people's home investment on rental basis only increased notably in 2014 and 2017, whereas that of freehold basis fluctuated over the years with its maximum in 2011 and minimum in 2018. As earlier stated in Table 7.3, there was more percentage increase in the income of investment B (Freehold Basis) than that of investment A (rental basis), and this can be linked to the active nature of the net income of old people's home investment on freehold basis, especially from other services carried out in these properties such as nursing home, care for the dementia etc. In addition, the peak in the investment A in 2017 is lower than that of the 2014, and this suggests that ROI of investment A declined over the years. Hence, Investment on freehold basis (B) seems to have a better prospect of ROI than rental basis. However, in terms of volatility, investment A shows a lower level of risk than investment B.

Net Present value, Internal rate of Return and Profitability Index analyses were conducted on investment "A" (which is on rental basis where the investor rented the property and provided facilities at the inception of the business) and Investment "B" which is on freehold basis occupying a purpose built facility for the investment. The result of the analyses using three discounting viability appraisal techniques and net cash flows is shown in Tables 7.4 and 7.5; Fig. 7.3.

Based on the decision rules of NPV, IRR and Profitability Index (PI) for independent projects, Table 7.5 shows that the two investments A and B (investments in Old Peoples' homes on Rental and Freehold basis respectively) are viable and acceptable

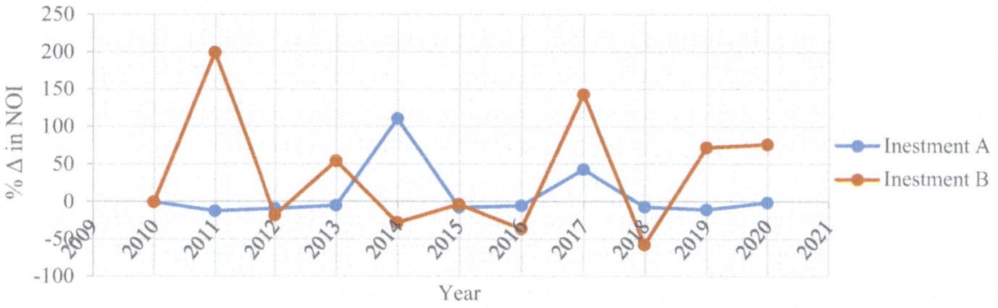

Fig. 7.2 Percentage change in the net income of old peoples' home investments

Table 7.4 Net present values (NPVs) of old peoples' home investments

Rate	18%	20%	25%	30%	40%	50%
NPV Invt **A**	15,398,058	13,088,636	8,471,319	5,066,651	514,708	– 2,291,626
NPV Invt **B**	14,760,150	3,992,463	– 17,711,159	– 33,927,178	–	–

Source Field Survey (2021)

Table 7.5 Investment viability using various appraisal techniques

Appraisal techniques	Investment A (rental basis)	Investment B (freehold basis)
Net present value (NPV) @ 18% target rate of return	15,398,058	14,760,150
Internal rate of return (IRR) (%)	41.53	20.81
Profitability index (PI)	2.23	1.12

Source Field Survey (2021)

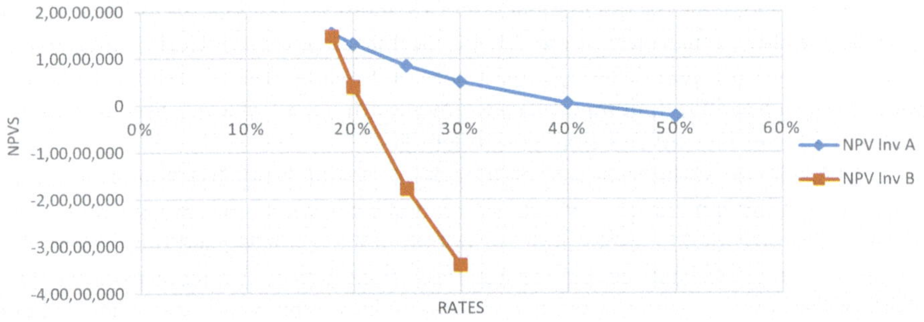

Fig. 7.3 Graphical determination of IRRs of investments on leasehold (A) and freehold (B) bases

with positive NPV greater than zero at 18% target rate, IRR above the target rate of 18% and PI greater than 1.0. However, investment A has a higher positive NPV, IRR and PI value than Investment B, thus, a better investment option than B. Furthermore, Fig. 7.3 shows that Investment B, which is investment on freehold basis is highly sensitive to little change in the rate of return, thus signifying very high riskiness of such investment. Hence, a risk averse investor would prefer investment in a less risky investment than that which has high level of variability in income.

There are factors influencing old peoples' home investment in Lagos, and these have been structured on a 5-point Likert scale where individual factors were ranked based on their weighted mean scores (WMS). This is presented in Table 7.6.

Table 7.6 shows that factors such as affordability, cultural and governmental disposition ranked very high among the factors that influence investment in Old Peoples' Homes in Lagos, with mean scores of 4.71, 4.50 and 4.25 respectively, while quality

Table 7.6 Factors influencing old people's home as an investment in Lagos State, Nigeria

Factors	SA (5)	A (4)	U (3)	D (2)	SD (1)	WMS	Rank
Affordability: people willing to but cannot afford the expenses	3(75.0)	1(25.0)	0.00	0.00	0.00	4.71	1**
Governmental: the lackadaisical attitude of the government towards providing working policies to encourage the investors and provision of enabling environment for such investment to thrive	2(50.0)	2(50.0)	0.00	0.00	0.00	4.50	2**
Cultural: the cultural belief that relatives of old persons should be responsible for their health	1(25.0)	3(75.0)	0.00	0.00	0.00	4.25	3**
Misconception: about old people's home as a residents for sick, smelling and insane elders	1(25.0)	1(25.0)	1(25.0)	1(25.0)	0.00	3.50	4*
Retirement issue: the older persons struggling to fit in to new way of life after many years of being occupied at work and the issues related to unpaid pension	0.00	2(50.0)	1(25.0)	1(25.0)	0.00	3.25	5
Financial: old people and their relatives in Nigeria are not socially and economically secured as a result poverty	0.00	1(25.0)	1(25.0)	1(25.0)	1(25.0)	2.50	6
Legal: regulations and restriction by the government that poses a threat to the investors return e.g. Tax	0.00	0.00	2(50.0)	1(25.0)	1(25.0)	2.25	7
Management and maintenance: the management culture and maintenance of the old people's home facilities for better service	0.00	0.00	1(25.0)	1(25.0)	2(50.3)	2.00	8
Quality of care: the quality of care provided by old people's home	0.00	0.00	0.00	2(50.0)	1(25.0)	1.50	9
Social: the problem of being able to relate with immediate environment or new mate in an old peoples facility	0.00	0.00	0.00	1(25.0)	3(75.0)	1.25	10
Competition: organizations providing the same service	0.00	0.00	0.00	0.00	0.00	1.00	11

Source Field Survey (2021)
SA = Strongly Agreed, a = Agreed, U = Undecided, D = Disagree, SD = Strongly Disagreed

of care, social factors and competition from other investors in the same industry ranked the least with mean scores of 1.50, 1.25 and 1.00 respectively. This implies that respondent strongly agree that affordability, Governmental and cultural factors affect old people's home as an investment in Lagos, while the respondent strongly disagree that factors such as quality of care, social factor and competition affect old people's home investment in the study area. This could be because the market is still new and very few investors in that market. The result further reveals that respondents are indifferent on whether issues on retirement affect the investment with the mean score of 3.25 and respondent also disagree that financial factors, legal issues, maintenance affect the investment in the study area. This study supports the work of Yang (2017) that affordability, cultural and governmental factors are major factors affecting Old' people's Home investment, but quite different from the work of Forder and Allan (2011) which concluded that factors such as management and maintenance of the elderly home, quality of care and competition are factors majorly affecting investment in Old people's home. Another reason why Affordability could be a major factor influencing the investment is that investing in Old people's home is capital intensive. This comes from high cost of infrastructural facilities used in the homes. Also, rate of taxation imposed on Lagos old people's home is enormous, thus, contributing to huge capital outlay and making it difficult for small investors to break into the market as well as not affordable to low income earners who cannot afford to pay for their services.

7.5 Conclusion

This paper has examined the viability of investing in old people's home in Lagos state, Nigeria. The study assessed the level of investment in old people's home in Lagos, evaluated the viability of old people's home investment and examined the factors influencing the investment in old people's home in the study area. Using descriptive statistics and three discounting appraisal techniques (NPV, IRR and PI), the result shows that there are varieties of services rendered by the old people's home. The Care home was a basic function of old people's home, while other services such as nursing home, adult home, home care with dementia, retirement home and assisted living home were additional functions or services that majority of old peoples home incorporate into their services. Furthermore, investment in Old people's home, whether on rental basis or freehold basis, was found to be viable on NPV, PI and IRR decision criteria. Finally, the result revealed that factors such as affordability, cultural dispositions and government's attitudes are major factors affecting old peoples' home investment in Lagos. Therefore, government should make policies that support and promote the growth of the industry and encourage prospective investors to take advantage of the market. More also, public enlightenment programs should be carried out on the benefit of adopting old peoples home care for their older persons. In addition, prospective investors are advised to take advantage of this novel and profitable investment thereby increasing care services to the masses as well as promoting the overall growth of the economy. Finally, further research could focus on the willingness of people to pay for the services rendered by old people's home in Lagos State, Nigeria.

References

Aboderin I (2004) Decline in material family support for older people in urban Ghana, Africa: understanding processes and causes of change. J Gerontol Soc Sci

Aboderin I (2016) Emerging research focus on long term care of older person in Sub-Saharan Africa. African regional conference on gerontology and geriatrics, Kenya

AgeUK (2020) Type of care home. Tavistock Square, London: Tavis House, pp 1–6

Ajomale O (2007) Country report: ageing in Nigeria—current state, social and economic implication. In: Hoff A (ed) Summer newsletter 2007 of the research committee (RC11) on the sociology of ageing of the international sociological association (ISA). Oxford institute of Ageing, Oxford, pp 15–20

Akinny WD (2016) Investigating the desirability and feasibility of the 'Old People's Home' as a viable business in Ghana. Ashesi University College, Department of Business Administration Unpublished

Allard T (2016) Nursing home profits soar as patient care declines. Retrieved from The Sydney Morning Herald: http://www.smh.com.au/federalpolitics/political-news/nursing-home-profits-soar-as-patient-care-declines20151224-glupug.html

Apere P (2015) Key challenges of Nigerian Pension Industry and possible solutions (1). The Nation. Retrieved 3 Sep 2020, from http://thenationonlineng.net/key-challenges-of-nigerian-pensionindustry-and-possible-solutions-1/

Cumming E, Henri WE (1961) Growing old: the process of disengagement. Basic Books, New York

Forder J, Allan S (2011) Competition in the care homes market. A report for the OHE

Fukui S, Yoshiuchi K, Fujita J, Ikezak S (2014) Determinants of financial performance of home-visit nursing agencies in Japan. BMC Health Serv Res 14(11):1–8

Imamoğlu C, Imamoğlu EO (2006) Relationship between familiarity, attitudes and preferences: assisted living facilities as compared to nursing homes. Soc Ind Res 235–254

Israel GD (2002) Sampling the evidence of extension program impact. IFAS, University of Florida, Program Evaluation and Organizational Development

Kehinde MO, Shittu AM, Adewuyi SA, Osunsina IOO, Adeyonu AG (2021) Land tenure and property rights, and household food security among rice farmers in Northern Nigeria. Heliyon 7(2):106–110

Laing, Buisson (2013) Care of elderly people UK market survey, 2012/13, London

Li D, Ban X (2015) Analysis of demands for community services among empty nesters in urban China. Bus Econ 4:21–54

Liu M (2010) Retirement community: a promising pattern for aging empty nesters in rural China. Adm Law 1:49–53

Nyanguru A (1990) The quality of life of the elderly living in institutions. Anchor Books, Gardwn city Newyork

Odaman MO, Ibiezugbe MI (2014) Health seeking behavior among the elderly in Edo

Okoye U (2004) The erosion of traditional forms of care for the elderly and its implication for the elderly in Nigeria. In: Conference proceedings of the African conference on aging: rapid ageing and the changing role of the elderly in African households, pp 224–236

Okoye UO (2013) Community-based care for home bound elderly persons in Nigeria; a policy Option. Int J Innov Res Sci Eng Technol 2(12):7086–7089

Oladeji D (2011) Family care, social services, and living arrangements factors influencing psychosocial wellbeing of elderly from selected households in Ibadan, Nigeria. Educ Res Int 1–6

Osunderu O, Adeyoyin AE (2018) Elderly abuse and care of the aged in Nigerian Society: a case study old people's home Yaba. Gerontol Geriatr Stud 2(1):1–6

Oyinlola O, Oyinlola TO (2017) Concept, conception and misconception of old people's home in Nigeria. MOJ J Gerontol 2(6):1–4

Population Reference Bureau (2012) World population data sheet. Author, Washington, DC

Róin A (2015) The multifaceted notion of home: exploring the meaning of home among elderly people living in the Faroe Islands. J Rural Stud 39:22–31

Reilly FK, Brown KC (2006) Investment analysis and portfolio management. South-Western, Cengage Learning: Natorp Boulevard Mason

Franchise Business Review FB (2012) Special report: Senior Care Franchises

Tanyi PL, André P, Mbah P (2018) Care of the elderly in Nigeria: implications for policy. Cogent Soc Sci 4(1):155–2016

Ugwu L, Ugwu D (2007) Attitudes of University employees towards ageing. ESUT J Psychol Stud 2:118–128

United Nations Department of Economics and New York Special Affairs Populations Division (2006) World population prospects. United Nations, New York

Wang L, Shi Y (2008) The mental health of urban empty-nest elderly. Chin J Gerontol 7:704

Yang S (2017) Influence factors of the demand of joining senior housing in Guangzhou, China: An analysis based on Guangzhou senior housing market. Instituto Universitaro de Lisboa, unpublished, pp 1–156

Chapter 8
Highest and Best Use Analysis of Real Estate Investment in Selected Cities in South Africa and Nigeria

Ijasan Kolawole and Adeyemo Adebola Olufunmilola

Abstract Investors and property developers are dealing with increased vacancy because of the recent COVID-19 pandemic, which primarily hit office buildings and commercial centers in urban areas. Vacancies and increased abandonment of particular property classes are increasing at an alarming rate in both Nigeria and South Africa. The national vacancy rate, according to the South African Property Owners Association (SAPOA) in 2022 was 11.6%, with Johannesburg having the highest vacancy rate at 15.8% in the first quarter of 2021 and 19.4% in the first quarter of 2022. In Ibadan, South-West Nigeria, 20% of residential houses developed in the last 50 years remain vacant. As a result, investors and property developers are constantly looking for new uses for unoccupied properties such as office buildings and commercial spaces. The highest and best use (HBU) analysis was described in this research to solve the vacancy problem in South Africa and Nigeria. The physically achievable, appropriately supported, and economically feasible use of vacant land or improved property to produce the most benefits as well as the highest land value at any given time. This study investigates how investors and property developers might utilize the concept of HBU analysis to remodel buildings and commercial spaces to ensure that they are not only profitable but also provide optimal value to cushion the aftermath of the recent global pandemic. The paper's key question is, "What are the best alternative uses for vacant property in the selected cities?".

Keywords Highest and best use · Property · Vacant spaces · Value

I. Kolawole (✉)
Black Research Network, University of Toronto, Toronto, Canada
e-mail: kola.ijasan@utoronto.ca

A. A. Olufunmilola
Department of Estate Management, University of Ibadan, Ibadan, Nigeria

S. P. Akinbogun et al. (eds.), *Facility Management practices*,
https://doi.org/10.1007/978-3-031-65321-6_8

8.1 Introduction

In real estate, highest and best use analysis is used to determine the most profitable and competitively productive use that a property can be subjected to (Rattanaprichavej 2014). This analysis utilizes four key metrics to come up with the appropriate use of a property. They are (1) physical possibility, (2) legal permissibility, (3) financial feasibility, and (4) maximum productivity (Rymarzak et al. 2022). This paper presents the highest and best use (HBU) analysis focusing on commercial real estate property in Johannesburg, South Africa, and the cities of Ibadan and Lagos in Nigeria. Lagos is be used as the case study for the vacant office spaces in Nigeria, while the vacant households will be discussed as observed in the city of Ibadan. Following the discussion of the broad influence of COVID-19 on the built environment, the first section concentrates on office spaces and other commercial property spaces that occurred as a result of vacancies created by the COVID-19 (coronavirus) epidemic in Johannesburg, South Africa. The analysis' main goal is to determine the best alternative use for commercial property in Johannesburg that has been affected by vacancy because of the recent COVID-19. Additionally, the second segment examines the Nigerian case by focusing on the cities of Lagos and Ibadan. The Lagos case focuses on office space issues, whilst the Ibadan case focuses on residential buildings. To contextualise the research, the rest of the paper starts from an overview of the built and human environment and how COVID affected city dwelling, next the concept of the highest and best alternative use analysis of real estate is discussed before delving into the focus cities of Johannesburg, Lagos and Ibadan.

8.2 The Impact of COVID-19 in the Built and Human Environment

The years 2019 and 2020 were marred by changes in the world as countries struggled to control the coronavirus pandemic, with the virus spreading fast, causing morbidities and fatalities of diverse degrees in different countries across different regions (Verma and Prakash 2020). The initial case of COVID-19 reported in Wuhan, China, back in November 2019 marked the beginning of a virus that spread across the world, including South Africa (Sekyere et al. 2021; Lee 2020). The country reported its first case on March 5, 2020, but an exponential increase in the early days to 61 cases by March 15, 2020, prompted President Cyril Ramaphosa to declare a national state of disaster, which was followed by a raft of measures to curtail the fast-spreading virus (Sekyere et al. 2021). The impact of COVID-19 on most African societies can be addressed in the social, health, environmental, technological, and economic spheres, as discussed in the next section.

8.2.1 Social Implications of COVID-19

The raft of measures commissioned by the government, including mandatory lockdown, effectively restricted people's movement in the country except for those working in essential services such as health, transportation, supermarkets, pharmacies, the fire department, law enforcement, and informal food trading, among others (Spaull and Van der Berg 2020). All other people were expected to stay indoors except in cases in which they needed to buy food, buy medication, or seek healthcare services (Saladino et al. 2020). The open outlets also had restrictions to ensure that no more than 50 people gathered in a place at a single time (Sekyere et al. 2021). Social-cultural functions, schooling, and religious gatherings were canceled owing to the social distancing policy that took effect at the time. Even the homeless people had to be relocated to well-spaced temporary shelters (Hedding et al. 2020). While the social distancing policies minimized the spread of the virus, they disrupted the social-cultural fabric of South African society. People could not meet or gather, as they often did, which negatively impacted their routines. They could not visit their families or relatives, leading to mental health problems, particularly for students who were out in colleges or universities and could not reach their families after the lockdown was imposed (Sekyere et al. 2021).

The low-income neighborhoods in South Africa and Nigeria are both informal, peri-urban, or rural dwellings characterized by closely located shacks (Nguse and Wassenaar 2021). These dwellings lack laid-out infrastructure, making it difficult to achieve social distancing. Overcrowded shacks can make it unhealthy for people to stay indoors for long periods of time (Sekyere et al. 2021). During COVID-19, the populations in these places went outside but met police and other law enforcement agencies on patrol, who forced them indoors, leading to growing tension between civilians and the police. The sale of alcohol as well as its consumption decreased considerably during the lockdown and social distancing measures (Sekyere et al. 2021). Consequently, social vices that come with alcohol consumption, including crime and addiction, decreased considerably at the time.

8.2.2 Health and Environmental Implications

The increase in reported cases of infections in South Africa piled pressure on existing healthcare facilities. The government established testing centers to ensure adequate testing of the populations (Nguse and Wassenaar 2021). There weren't as many cases as possible of COVID-19 in Nigeria, and different theories have been postulated for this. The reasons given ranged from a lack of testing (Chitungo et al. 2020) to the young age of the population and other factors such as the limited prevalence of old people's homes. The government also established isolation centers to quarantine those infected by the virus. The government recommended health and environmental measures such as practicing good personal hygiene, avoiding body contact and handshakes, staying indoors, staying hydrated, and frequently washing hands (Naidu 2020). The government also required facilities such as hotels and restaurants to be sanitized and cleaned to avoid the spreading of the virus through contact with surfaces.

8.2.3 Technological Implications

The pandemic showed how technology can be used both beneficially and negatively during crises. Social media, although helping to spread information, led to anxiety and misinformation about the pandemic (Broadbent et al. 2020). On a positive note, the use of contact tracing through technology helped track the chain of infections, which helped curb the spread of the virus. The use of Zoom and other modern information technology empowered people with the capability of working from home (Rogerson and Rogerson 2020). Although this was beneficial in controlling the virus, it culminated in the loss of office jobs and the vacancy of office spaces in cities such as Johannesburg (Sekyere et al. 2021).

8.2.4 Economic Implications

The stay-at-home orders aggravated the already high poverty levels in South Africa owing to no economic activities and no income, mostly for the self-employed as well as those in the informal sector (Schröder et al. 2021). The transport sector is one of the worst affected because it cannot operate. With the country considered to be in a recession before the pandemic, the stay-at-home orders and lockdown implications piled further tremendous pressure on the country's formal productive sector (Sekyere et al. 2021).

8.3 Concept of the Highest and Best Alternative Use Analysis of Real Estate

Highest and best use (HBU) analysis is an economic concept used in appraising the market value of real property. The HBU of a property is defined as the physically achievable, appropriately supported, and economically feasible use of vacant land or improved property to produce the most benefits as well as the highest land value at any given time (Sudharma and Rahmadani 2020). HBU analysis is applied when appraising the value of property to determine the most profitable and feasible use.

Figure 8.1 shows the four criteria followed in HBU analysis, namely, legally permissible, physically possible, financially feasible, and maximally productive (Degli-Esposti 2019). The data to be used in such an analysis, however, is derived from a general market study, a marketability study, and a feasibility analysis. The general market study encompasses supply and demand analysis to determine the appropriate potential uses for property (Aleksandrovna 2019). The marketability study is a six-step process that forecasts the performance of property under specific conditions (current market conditions or predicted market conditions).

The feasibility analysis entails the use of market data to assess whether the potential uses of property are feasible (Degli-Esposti 2019). HBU analysis is the final stage that examines the alternative uses of property that can achieve the highest value. The analysis follows four-step criteria based on the pyramid in Figs. 8.1 and 8.2.

Fig. 8.1 HBU criteria Pyramid (Google Stack 2023)

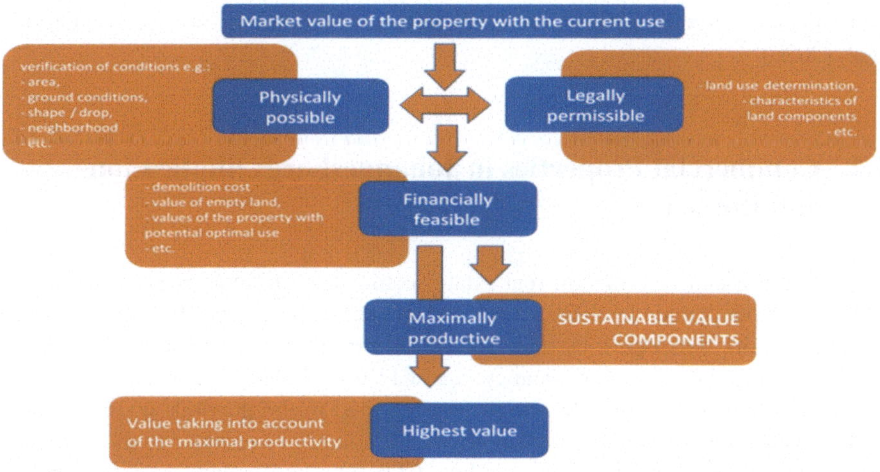

Fig. 8.2 Procedure of highest and best use analysis for the real estate valuation. *Source* Walacik et al. (2020)

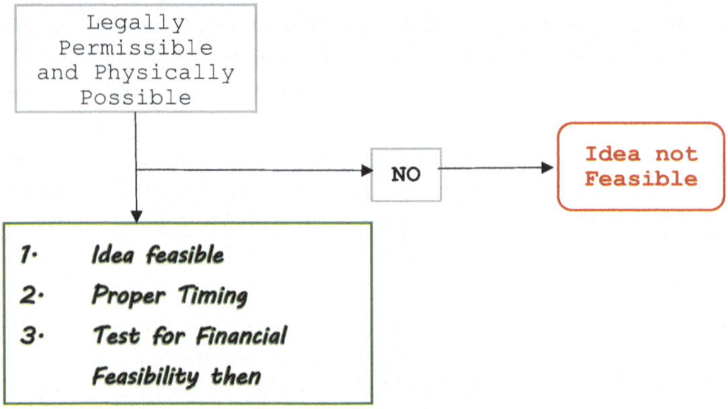

Fig. 8.3 Testing HBU options

The four criteria are followed in order, starting from the bottom of the pyramid in Fig. 8.1, and moving upward. Mostly, legal permissibility and physical possibility are the first steps and can be applied in any order (MAI 2015). They should be applied before moving on to financial feasibility. When the proposed use is financially feasible yet physically impossible or legally prohibited, the proposed use becomes irrelevant. Thus, the proposed use must be physically possible, legally permissible, and financially feasible before it can be tested for economic productivity (Aleksandrovna 2019). The flow chart in Fig. 8.3 shows a typical HBU analysis process.

Aside the physical deterioration caused by high vacancy rates in cities, financial sustainability challenges are additional issues for property owners. There are however a range of alternative uses for commercial properties that have been covered in literature, and some have been practically utilized and found to be effective; Drgoňa et al. (2020).

8.4 Commercial Properties in Johannesburg: Implications and Impact

The South African commercial real estate sector faced difficulties even before the COVID-19 pandemic. In 2020, when the pandemic was just starting, an office vacancy report developed by the South African Property Owners Association (SAPOA) placed the national vacancy rate at 11.6% (Walker 2021). The figure was considered higher than the 7% which was reported in the previous quarter, which SAPOA attributed to the fragile economy in the country as opposed to the influence of COVID-19 (Walker 2021). Nonetheless, COVID-19 propelled a process that had started in which work practices were changing from people working in offices to people working at home. COVID-19 worsened the situation for commercial property owners through "shadow vacancies," characterized by a surge in the number of vacant properties as many businesses that hitherto operated in rented office spaces shifted to the work-from-home (WFH) setup (Research and Markets 2023). The move saw many businesses downscale their physical offices for smaller offices that facilitated a hybrid workforce that easily worked from home. Johannesburg is the leading economic and financial hub in South Africa. It is where deals are made, where the money is. The general property valuation in the city as of 2023 shows an increase of 12% in the past five years to R1.59-trillion (Mkentane 2023). This fast-paced metropolis is the first stop for any big-thinking entrepreneurs (Quirk 2021). Johannesburg is the country's economic powerhouse, generating 17% of the nation's gross domestic product (GDP) mainly through the service, retail, and manufacturing sectors. Coca-Cola, Nokia, Zurich Re, and McDonald's—just to mention a few—have their head offices situated in the city of Johannesburg (Babalwa Soga 2022). The city hosts major banks as well as the largest stock exchange in Africa,—Johannesburg Stock Exchange in Sandton, making it a pinnacle financial sector in South Africa (Lukasiewicz 2020). The retail industry thrives in Johannesburg due to the relatively cheap rates in Africa, which makes it an attractive destination for retailers mainly from the neighboring countries in sub-Saharan Africa. The Sandton City precinct and the inner-city business districts are the two biggest retail precincts in Johannesburg (Lukasiewicz 2020).

Johannesburg has historically experienced robust office fundamentals, which eroded significantly following the COVID-19 pandemic. The vacancy rates steadily increased with WFH and corporate consolidations, resulting in more commercial property spaces being given up in the commercial property retail market (Deghi et al. 2022). Johannesburg recorded the highest office vacancy rate in the second quarter (Q2) of 2022 in South Africa's major metropolitan areas. It also experienced the highest office real estate vacancy rate in the first quarter (Q1) of 2022, increasing from 15.8% (Q1 of 2021) to 19.4% (Q1 of 2022) (Research and Markets 2023). COVID-19 played a huge role in the changes in vacancy rates and office demand dynamics given the move by corporates to scale down on space requirements to align with the new office rotational hybrid model and the WFH model. Consequently, the office foot count decreased considerably from pre-pandemic levels. The commercial property spaces previously used as offices for firms were left vacant, with COVID-19 presenting a serious challenge to both major and minor commercial real estate players such as Chas Everitt Property Group, Excellerate JHI, Pam Golding Properties, Amdec Group, and Growthpoint Properties (Research and Markets 2023). Having discussed and ascertained the physical impact of the pandemic on office spaces, i.e., the apparent reduction in need and demand, the next section discusses the potential for an alternative use of these excess spaces, both from the perspectives of the investors as well as the current leaseholders. It covers some of the highest and best alternative use ideas that property developers in Johannesburg can adopt. This is by no means an exhaustive list; however, it represents some of the adaptable options from international best practices and contextual local situations.

8.4.1 *Conversion of Office Buildings into Mixed-Use Buildings for Lifestyle Activities*

Landlords and investors struggling with high vacancy rates in the city may want to contemplate converting their office buildings into uses that are in higher demand within the location (Zhang 2020). For instance, Fairfax County, Virginia, in the United States, once struggled with 18 million square feet of unoccupied office space (Kirk 2019). This prompted a local planning workgroup to look for opportunities to recycle the old office buildings. They converted them into food incubators, co-working spaces, schools, and apartments. The recommendations were approved in 2015 by the county's Board of Supervisors with a caveat that the buildings be zoned for industrial or mixed-use purposes (Kirk 2019). The offices have since been used sustainably as apartments, live-work units, and schools. The aspect to consider in this situation is that the mixed-use purposes proposed were for activities on high demand. The residents in the area needed spaces for schools, apartments, and live-work units as the most demanded purposes (Kirk 2019). In real estate, the type of preferred activities in the surroundings of commercial properties typically determines what purpose they can serve (Lambourne 2022).

The office buildings in Johannesburg that were unoccupied during COVID-19 can use the aforementioned concept. For high-rise office buildings with different floors, individual floors can be utilized for different purposes depending on the in-demand socio-economic activities in the city (Hoesli and Malle 2022). For instance, one or two floors can be converted into mixed-use for lifestyle activities such as high-end fitness centers, high-end barber shops, high-end massage parlours, and high-end spas. At the height of COVID-19, it was difficult for people to visit gyms and exercise. More people are visiting gyms in South Africa to get back in shape now that the lockdown and other restrictions have been lifted (Wilke et al. 2021), but adding other uses to a building will further promote the best uses of buildings. Further, the economic revival after Covid-19 era has seen an increase in lifestyle activities, propelled by increased income among the middle-class and the high-class (Ramagole et al. 2020). Thus, setting up areas for lifestyle activities such as spas and massage parlors in high-end office spaces can be profitable. These lifestyle activities are interrelated in a sense. For instance, on one hand, when a customer visits a barber shop, they can have the impetus to try out the massage parlors or the spas (Armstrong 2020); on the other hand, workers in a building might start planning their days such that they kill two birds with one stone, whereby they go to work in the morning and try other activities within their building during the evening.

In terms of legal permissibility, the concerns that may emerge with mixed-use for lifestyle activities are zoning laws, building codes, environmental regulations, etc. Zoning laws would not be required to change because the proposed activities are more commercial than industrial (Scheba and Turok 2020). Thus, their ideal location is the city center. The building codes would not be much affected either because slight renovations that would not require major redesigns would be needed. However, environmental regulations would be problematic, especially for fitness centers, where noise may be a problem. This would require compliance with sound pollution requirements that would see the incorporation of soundproofing (Bushe 2019). In terms of physical possibilities, large office spaces can hold gyms, while other medium-sized offices can hold spas, massage parlours, and barber shops. Thus, the size would only be problematic where there are no large open spaces in the buildings for the gyms. Accessibility would not be an issue because the buildings are in the city center, which is frequented by people who are either going to work or doing their shopping (Brill and Robin 2020). Further, the city center features customers who shop in high-end shopping malls and can visit spas, gyms, barber shops, and massage parlours. The city is a high-traffic area that can attract different kinds of people, including those coming from outside the city (Brill and Robin 2020). Most of the commercial office buildings in Johannesburg have parking spaces. Therefore, motorists can easily access the building and enjoy the lifestyle activities provided (Yusuf et al. 2021).

With strong marketing, these spaces can be profitable, thereby ensuring financial feasibility. The capacity for profitability will depend on the number and type of customers. Marketing that can ensure the middle class and the high-class visit the new business can lead to returns that cover renovation and rental costs with an adequate profit margin (Akinsomi 2021). Maximum productivity will be ensured by the mixed-use approach adopted for the vacant office buildings. The location of lifestyle activity shops within the same building or flow can attract Gen Z (people born from 1997) customers of different kinds seeking different lifestyle services (McKay et al. 2021).

8.4.2 Conversion of Open Office Spaces into Food Stores

Healthy eating has become a lifestyle trend with an increase in lifestyle-related diseases (McKay et al. 2021). South Africans have not been left behind in terms of healthy eating. There is a high demand for healthy foods such as vegetables, groceries, and proteins in the country (Mbungo 2020). However, most people, especially in the city, are conscious of the food they eat for fear of food poisoning. Further, they are concerned about the nutritional content of the foods (Mbunge 2020). Food stores were ranked among the Top 20 Best Small Business Opportunities in Johannesburg by Profitable Venture Magazine (2021). Some of the notable ideas for food stores include groceries, butcheries, shopping centers, and restaurants. The people living in the city center would benefit from fresh foods that they would otherwise need to seek expensively from surrounding neighborhoods more accessible to farms (Wekeza and Sibanda 2019). Thus, fresh foods sold in grocery stores would be in high demand, besides packed meat. Shopping centers and restaurants would also attract demand, especially from people who come to shop in the grocery stores (Rahman et al. 2021). Various developers have successfully converted office buildings into hotels. The Hyatt Place at 1522 K Street in Washington, D.C., was a former office building redesigned by Cooper Carry into a hotel (Kirk 2019). A developer in Indianapolis converted an office building situated at 141 E. Washington St. into a $25 million Marriott-branded boutique hotel (Kirk 2019).

In terms of legal permissibility, the major worry would be the building codes. Because food stores are commercial establishments subject to existing zoning laws, they would also apply to the recommended usage (Iloms et al. 2020). However, the building codes would change in line with health requirements. This would require the property handlers to comply with building codes that ensure that the office spaces are safe and suitable for holding food stores (Bertaud 2021). In terms of physical possibility, the large office spaces are of the appropriate size to host medium- to large-sized food stores covering groceries, butcheries, and shops. The food stores would not require major redesigning or compartmentalization. All that would be needed are minor fittings and the interior design of the spaces to fit the use (Tokazhanov et al. 2020). For instance, a restaurant would require the arrangement of chairs and tables in the open space and the construction of a kitchen counter. The buildings are located in the city center, where they receive high traffic. Coupled with ample parking spaces, people looking for groceries and meat or those who want to use restaurants can visit the premises (Florida et al. 2021). The financial feasibility of the recommendation will depend on the marketing campaign, which can effectively attract an influx of customers within the city. At optimum, the food stores can generate enough income to break even and even generate income (Florida et al. 2021). Using a variety of food stores ensures maximum productivity. This is premised on the idea that not all the people coming into the building would want to buy groceries. Some of them would want to eat in restaurants.

8.4.3 Conversion into Art Exhibitions and Studios, Architectural Studios, and Engineering Design Offices

Vacant office buildings can have large office spaces that can be used as exhibitions and studios not only by artists but also by architects and engineers. Johannesburg is an economic hub that attracts artists, architects, and engineers, among other professionals. While COVID-19 triggered the work-from-home (WFH) setup that has increasingly been adopted by companies, some occupations require office spaces. For instance, artists involved in drawing artwork require spaces to do their work and to display it (Florida and Pedigo 2020). In South African culture, a range of artists display indigenous art pieces that can attract customers. Architects and engineers in Johannesburg working for local or international companies also require spaces to work out their designs, while engineers require spaces for their plans (Madeddu and Clifford 2021). The open spaces in the abandoned office building provide suitable space to display artwork as well as set up an art studio (Habelsberger and Bhansing 2021). The South African culture is rich in artistic elements that can be commercialized, especially to target high-end customers. These spaces can provide a link between the artwork and such customers. They can visit the buildings to experience the exhibitions and make their purchases (Pataki-Bittó and Kapusy 2021). Engineers and architects often locate their offices in city centers where they can get the appropriate clientele (Madeddu and Clifford 2021). Therefore, the vacant office buildings can be advertised as offices for engineering and architectural work.

In terms of legal permissibility, studios do not attract any significant problems because their use is almost like office use. Thus, the zoning laws, building codes, and environmental regulations that apply to offices would also apply to the proposed studios. In terms of physical possibility, the size and shape of the vacant offices can accommodate studios (Bonita and Wadley 2022). The spaces used for the offices can ideally be used for architecture and art studios. For art exhibition spaces, there would require open spaces that would allow the art pieces to be displayed in areas that would allow for the movement of customers (Ellen and Kazis 2022). The spaces would not require any significant renovation or interior design. For the art studios, the common modifications would be having display boards for the art and a workstation for the drawings. For the architecture studios and engineering design offices, huge tables and displays for the drawn designs will be needed alongside computers to form the workstations. All these do not require any major modifications to the structure of the offices (Remoy and van der Voordt 2014). However, they would need to be painted white because most studios have white as their preferred color, although other colors would also be used depending on the preference of the studio owners. The office buildings are accessible because they are in a high-traffic area (Ellen and Kazis 2022). Consequently, prospective clients can easily drive to the buildings and park in the available parking spaces.

In terms of financial feasibility, office buildings can attract huge rents from studios. Typically, the rental cost for architecture studios, engineering design offices, and art studios is considerably higher than the rates that would be paid for normal offices (Madeddu and Clifford 2021). Thus, the developers can derive considerable returns when they manage to attract clients needing offices for art studios, architecture studios, and engineering design offices. Using the vacant offices as art studios, architecture studios, and engineering design offices allows for maximum productivity (Mukherjee and Narang 2022). Such a mixed-use approach helps to draw different types of clients in the arts, architecture, and engineering fields. Thus, the developers do not only have to depend on one of these for rent. Further, the mixed-use approach allows the developers to negotiate for the best rates in each category. For instance, the rental rates for art studios may be significantly higher than the rates paid for engineering design offices (Mukherjee and Narang 2022). The mixed-use approach helps to attract different clients, thereby alleviating the shortage of clients.

8.4.4 Conversion into Air BnB and Co-living Apartments

Another sharing economy concept has developed in the modern generation: co-living. In cities such as Johannesburg, where the housing crisis and high rental costs are problematic, property developers have an additional investment opportunity through co-living (Simpson 2020). This is a concept in which people share their living costs. This idea has been tried in Johannesburg, where office blocks were converted into affordable housing in the inner city (Kotze and Carow 2019). Besides co-living, there is a trend towards Air BnB apartments, in which people live in fully furnished apartments for a fee and experience a home-like feeling while living in the apartments. The vacant office buildings in Johannesburg can be used for co-living and Air BnB. The open office spaces can be converted into co-living units with beds placed in the office spaces separated by simple partitions (Kotze and Carow 2019). This idea was utilized in London, where a start-up called the Collective converted its office building located in the city into the largest co-living building in the city. The Collective Old Oak, as it is known, has 550 beds that offer suitable accommodation for co-living (Kirk 2019). The walled offices in the vacant office buildings can be converted into Air BnB units, furnished to serve the purpose.

Concerning legal permissibility, the existing zoning laws would not provide any restrictions because co-living units and Air BnB units are structures that allow for temporary occupancy (Wilkinson and Remoy 2017). This relates to offices in which people occupy the building temporarily and not permanently. The building codes would not differ as well because the strength parameters during the design and construction take into consideration the weight that would emerge from people occupying the building (Government of South Africa 2023). However, the building codes would need to be reviewed to determine their suitability. For instance, the buildings must have safe lifts or stairs to allow people's movement. Further, isolated columns would be a safety concern. Occupancy issues are likely to arise in terms of the reasonable timeframe allowed to make the renovations (Government of South Africa 2023). Thus, adequate marketing should be done to determine the potential demand, which would make it easy to prepare for renovations in good time to avoid any inconveniences to tenants. The physical location of the building in the city center is strategic to attract both city dwellers and travelers looking to put up temporarily. Further, the buildings are in a high-traffic area, hence, strategically capable of attracting customers (Simpson 2020). People can also drive and park in the parking spaces. Having both co-living units and Air BnB units ensures maximum productivity by allowing prospective customers to choose their preference. Thus, the property developers can derive rent from both.

8.4.5 Conversion into Security and Courier/Delivery Services

There is a rapid increase in the rate of crime in major cities in South Africa, with Johannesburg being one of the high-risk security areas (Navsaria et al. 2021). This makes starting a security outfit a smart business idea. The Profitable Venture Magazine (2021) ranked security outfits among the top five smartest business ideas in Johannesburg. The business would sell due to the rapidly increasing crime rates in the country. Many firms in Johannesburg typically outsource their security services to third-party companies. Thus, the vacant offices can be used as offices for security firms. The offices can also be used as shops for the sale and installation of security devices such as CCTV cameras. Most security firms also provide courier/delivery services. The courier business is also a thriving one. COVID-19 opened the doors for courier companies, and now the delivery trend has reached its peak with people opting to make purchases and having them delivered (Navsaria et al. 2021). The vacant offices can be used as offices for courier service firms. Besides, the increased crime rates in Johannesburg create a need for commercial storage spaces. Thus, businesspeople or other individuals in the city can use the office buildings as storage for their items (Marutlulle 2021).

The zoning laws, building codes, and environmental laws that apply to office buildings would not be different from the ones that apply to courier and security firms. They are just the laws and codes that apply to any typical office (Wilkinson and Remoy 2017) not be different from the ones that apply to courier and security firms. They are just the laws and codes that apply to any typical office (Wilkinson and Remoy 2017). Therefore, no changes or new compliance requirements would be needed for firms seeking to operate security firms and/or courier services (Government of South Africa 2023). In terms of physical possibility, the size and shape of the vacant offices would be ideal for security firms and/or courier services. The physical location is in the city center in a high-traffic area; hence, the potential demand for the services is high. The security firms and courier services would pay consistent and sustainable rent because they often set up permanently in a location as opposed to temporarily (Todes et al. 2020). Finally, maximum productivity would be ensured by having the vacant offices used as security firm offices, courier service offices, or both. This ensures that the demand for the offices is not only determined by security but also by the need for delivery. These purposes align with the need to solve the security problem in Johannesburg and the need to provide delivery services to city residents in line with the growing modern trend toward courier services (Braithwaite 2020). Having presented and discussed some of the HBU possibilities for Johannesburg, the next section presents the similar cases of two Nigerian cities and the HBU suggestions.

8.5 Commercial and Residential Properties in South-West Nigeria: Implications and Impact

In Nigeria, the major effect of the COVID-19 crisis on Nigeria has been economic (Obi et al. 2020). The pandemic ushered in Nigeria's deepest recession since the 1980s, with the greatest effect on services and industry. The resultant restrictions and lockdowns also created all kinds of headaches for the property sector in Nigeria. There was a significant drop in real estate transactions (Keke et al. 2020). This is because the rules regarding social interactions and isolation made it harder to arrange viewings and inspections, and many transactions were put on hold with the massive shut-down of construction sites. A new norm emerged for office spaces as many businesses now operate remotely. The 'work from home' experience established that businesses could operate with smaller but functional office spaces, especially in the service sector. The reduced need for physical discussions resulted in a reduced demand for office spaces, and residential apartments with comfortable spaces have been converted to serve the dual purpose of work and home.

About 20% of residential properties became vacant within the last two decades in Ibadan, the largest city in Nigeria (Falola 2023). These residential properties, most of which are found in the old government reserved areas (GRA) like Bodija in Ibadan comprise large family houses of five or six bedrooms, and the owners have become old, some are deceased, and many of their children have relocated outside the country. For those who are still alive and have not relocated among the owners of these gigantic structures, many of them have found the need to move to smaller bungalows that require lower maintenance. The smaller houses now accommodate the old couple, whose children have moved to start their own families. The issue of health is another reason for moving into bungalows. At their age, many cannot keep up with the daily ordeal of climbing stairs within the houses that were mostly designed to have the master bedroom upstairs.

A major reason for the void state of these properties is their non –affordability. Sixty percent of Ibadan working class dwellers are low or no-income earners (Babatunde et al. 2020). About 70% of houses the GRAS are large duplexes, detached and semi-detached houses, built about three decades ago, where the elites that could afford to build such houses lived within the city (Makinde 2020). The younger generation of homeowners now build smaller houses in more modern designs with functionality and cost as important considerations rather than prestige or status, which were important considerations to the older generation. According to Adeleke (2019), majority of families in Ibadan are tenants and about 44% will do with a three-bedroom flat. As such, there is no justification to want to rent a five- or six-bedroom mansion, purpose built for the older generation, and not easy to maintain. The last decade has witnessed the rapid pulling down and conversion of some of these vacant properties into terrace houses, club houses, and hotels. The last decade has witnessed the rapid pulling down and conversion of some of these vacant properties into terrace houses, club houses, and hotels. Lagos and Ibadan will be discussed as cases studied for these vacant offices and residential properties respectively.

8.6 Office Spaces in Lagos: Highest and Best Use

Lagos is the leading commercial nerve center of Nigeria and the city has witnessed rapid conversion of traditional core residential and high-density residential uses into commercial uses (Ankeli et al. 2021). Lagos is presently characterized by commercial ribbon street development, such that virtually all high-density areas are in a chaotic state with an indiscriminate mix of commercial, light industry, transport, and religious land uses. This paper recommends several conversions for vacant office spaces, bearing in mind the concept of highest and best use.

8.6.1 Conversion of Office Buildings into Co-working Spaces for Service Providers

Iroham et al. (2022) affirmed the emergence of co-working spaces as a viable use of commercial spaces in Lagos, Nigeria. Co-working is an arrangement in which workers for different companies share an office space. It allows cost savings and convenience using common infrastructure, such as equipment, utilities, conveniences, and a receptionist. Since only essential duties require a physical presence, many small businesses, especially in the service industry, will do with little and more affordable space. Since Nigeria still has a peculiar lingering problem of erratic power and limited access to the internet due to high costs, companies will not want such excuses from workers whose homes cannot provide power and internet access. They now coexist where such facilities are already provided in commercial areas. Many sole proprietors and small-scale businesses have embraced the use of co-working spaces in Nigeria (Adegbiji 2021). This is physically feasible and more economically viable, as large unused vacant spaces can now be partitioned and converted into different and smaller sizes without any legal implications.

8.6.2 Conversion of Office Buildings into Mixed-Use Buildings

Many large offices that previously hosted one company are fast becoming shopping malls that host legal firms, estate surveying firms, dental clinics, travel agencies, insurance brokers, and shops such as boutiques, household goods store, grocery stores and in recent times, even considered for accommodation (Ugwuonah 2022). This is an emerging scenario not only in Lagos but in many state capitals in Nigeria. Here, minimal conversion is necessary, which is legal, feasible, and viable.

8.6.3 Conversion of Open Office Spaces into Food Stores and Courier Delivery Offices

An increase in space demand for warehouse and retail space is beginning to spring up as pick-up offices for online shopping activities, including groceries, using different e-commerce applications (Zumstein et al. 2021). According to Abiamuwe et al. (2022), Nigerians have been found to prefer goods delivered to them from online purchases using courier services. Vacant commercial spaces in Lagos will do well if converted pick up shops or courier offices for the Jiji, Konga, Jumia amongst other online platforms.

8.6.4 *C*onversion of Old Structures or Unused Commercial Property into Car Parks

Dilapidated buildings exist in the central business district and attract so little revenue, even though they sit on prime land. Such will continue to experience voids (Orekan and Adebayo 2020). These old and vacant properties will be better pulled down and used as car parking spaces. The importance of parking spaces in some unplanned and crowded Nigerian commercial districts cannot be overemphasized.

8.6.5 Conversion of Vacant Multi-Story Buildings into Complementary Uses; Restaurants, Gyms, Games Arcades, and Worship Centers

Long duration of time at work have been observed amongst career couples in Lagos (Owolabi and Ajibose 2019). Conversion of vacant commercial spaces to relaxation centers that can provide food, relaxing sports, and games will be lucrative in commercial districts. Churches also can capture members who will not attend midweek services unless they are near their place of work. In this way, they stay in their work environment to attend service when the Lagos traffic has cleared and then go home within a short time. This is physically possible due to the fact that not too many large spaces are required to accommodate branches of churches. As such, there will be minimal conversion. Meetings are mostly held after work hours, and as such, there are more spaces to accommodate the cars of members. Also, there are no legal issues as churches are preferred in commercial areas over gated residential neighborhoods.

8.7 HBU for Vacant Household Properties in Ibadan, Southwestern Nigeria

This paper also examines the highest and best use of these vacant Ibadan household properties, bearing in mind the four criteria of HBU analysis, namely: legally permissiveness, physical possibility, financial feasibility, and maximum productivity.

8.7.1 Conversion into Air BnB, and Co-living Apartments

The african culture, unlike the western world, reflects communal living in which larger spaces are required. According to Issaka et al. (2022), Air BnB has been observed in Ghana to be some form of threat to the traditional hotel arrangement. Families who come for weddings in cities where they do not reside may find it easier to get bigger houses at more affordable rates than hotels can offer. On the issue of legal permissibility, GRAs in Nigeria are, in recent times, hosts of hotels and guest houses. As such, with little renovation, these structures can be modified for more productive and lucrative Air BnB use.

8.7.2 Conversion to Condominiums

Many would rather live nearer the central business district in Ibadan (Taiwo 2022). This ensures proximity to work. Because rental values in CBDs are higher than those in farther distances, these large vacant *structures* can be modified to accommodate studio apartments, single rooms, and two- and three-bedroom apartments all in the same property. These will be far cheaper, and they can be more easily rented than when you must look for a single tenant for the whole building. Condominiums are not different from other rented apartments in the neighborhood, and as such, there will be no legal issues. In this converted state, the properties are put to more lucrative use, and the void is reduced considerably.

8.7.3 Private Schools and Creches

Private schools and creches will be the most physically possible use of vacant properties in residential areas because only little modifications may be required; rooms are easily converted to classrooms, while the living rooms can be used as staffrooms. Most families in Nigeria consider the proximity of their ward's school to their home so they can find their way home after school. The reason is that most have to work late. For those who have toddlers, creches also offer afterschool care, where they clean, feed, and put the babies to sleep with all the facilities of a home in place before the parents return and pick up the children on their way from work. This is legally permissible and economically feasible. There is an emerging scenario of private schools springing up within residential areas that are quickly accommodating mixed land uses.

8.7.4 Churches

The widely known Redeem Christian Church of God and some other Pentecostal denominations will find some of the vacant properties useful. This used to raise legal flags in times past because of the noise emanating from the night vigils. Some churches in residential areas have held evening meetings rather than vigils. This is a feasible and productive rental use now that most denominations want to have branches that are closer to their members.

8.8 Conclusion

This chapter provides investors and property developers in the case cities with an overview of some of the HBU case studies that can be applied to the empty office buildings and commercial spaces caused by the recent COVID-19 outbreak. The chapter was prompted by the question "what are the best alternative uses for vacant property in the African cities?" with the examples of Johannesburg, Lagos, and Ibadan.

Based on an examination of the theoretical framework of highest and best use analysis of real estate, it has become clear that the dynamics of various cities vary, and that the possibilities for redevelopment or conversion are the primary factors in determining the optimal solution. In the light of this, it is possible to draw conclusions about potential avenues for growth, such as the opening of fitness centers, spas, massage parlors, and barber shops in former office building spaces. It is possible to transform the open workplaces into retail food establishments like supermarkets and eateries. They are versatile enough to serve as spaces for creating and displaying artwork, designing buildings, or even designing machines. Johannesburg's housing issue can be alleviated by converting the structures into Air BnB accommodations or co-living apartments, the overbearing residential properties in Ibadan could be redesigned to accommodate smaller family sizes and co-working is a potential for Lagos office market in order to prevent an oversupply. It is also suggested that spaces can be opened up for usage by security firms and delivery services. The legality, plausibility, cost-effectiveness, and potential output of these HBUs are reiterated as being of utmost importance.

References

Abiamuwe NO, David OR, Seriki-Mosadolorun JS, Olusegun RA, Nwahunanya I (2022) E-marketing of clothing products and consumers' satisfaction of delivery methods in Lagos State, Nigeria. Int J Home Econ Hospitality Allied Res 1(1):60–68

Adegbiji AP (2021) Technology entrepreneurship and innovation development in Nigeria: a study of innovation hubs in lagos state. In: International conference on innovative systems for digital economy| ISDE, p 223

Adeleke FG (2019) The dynamics of housing preferences and affordability in Oyo State, Nigeria. Ibadan Plann J 8(1 & 2):10–25

Akinsomi O (2021) How resilient are REITs to a pandemic? The COVID-19 effect. J Property Investment Finance 39(1):19–24

Aleksandrovna OA (2019) Real estate appraisal methods with the highest and best use concept. Available at https://dspace.spbu.ru/bitstream/11701/26145/1/st062407.pdf. Accessed on 13–06–2023

Ankeli AI, Nuhu MB, Popoola NI, Kemiki OA, Okoh SO, Omotehinshe JO (2021) Impact of COVID-19 pandemic on tenancy agreement: the lesson from an emerging rental housing market in Nigeria. Baltic J Real Estate Econ Constr Manag 9(1):79–93

Armstrong G (2020) The adaptive reuse predicament: an investigation into whether building regulation is a key barrier to adaptive reuse of vacant office buildings, Doctoral dissertation

Babalwa Soga SVD (2022) Significance of E-government services for small businesses in the City of Johannesburg. J Southwest Jiaotong Univ 57(4)
Babatunde O, Awofodu JO, Oladunmoye OMA (2020) Critique of housing affordability: appraising factors affecting no/low income earners in Ibadan
Bertaud A (2021) Order without design: how markets shape cities. Town Reg Plann 79:2–5
Bonita R, Wadley D (2022) Disposal of government offices in Jakarta pending relocation of the Indonesian capital: an application of multi-criteria analysis. Property Manag
Braithwaite J (2020) Crime as a cascade phenomenon. Int J Comp Appl Crim Just 44(3):137–169
Brill FN, Robin E (2020) The risky business of real estate developers: network building and risk mitigation in London and Johannesburg. Urban Geogr 41(1):36–54
Broadbent A, Combrink H, Smart B (2020) COVID-19 in South Africa. Glob Epidemiol 2:100034
Bushe B (2019) The causes and impact of business failure among small to micro and medium enterprises in South Africa. Africa's Public Service Delivery Perform Rev 7(1):1–26
Chitungo I, Dzobo M, Hlongwa M, Dzinamarira T (2020) COVID-19: unpacking the low number of cases in Africa. Publ Health Pract 1:100038
Deghi A, Natalucci MFM, Qureshi MS (2022) Commercial real estate prices during COVID-19: what is driving the divergence?. International Monetary Fund
Degli Esposti F (2019) The highest and best use analysis as a tool to determine the value of real estate asset: application to the former Philips headquarter in Monza
Drgoňa J, Arroyo J, Figueroa IC, Blum D, Arendt K, Kim D, Ollé EP, Oravec J, Wetter M, Vrabie DL, Helsen L (2020) All you need to know about model predictive control for buildings. Ann Rev Control 50:190–232
Ellen IG, Kazis N (2022) Flexibility and conversions in New York City's housing stock: building for an era of rapid change. NYU Furman Center
Falola T (2023) Ibadan in Ibadan: the university and the city. Yoruba Stud Rev 8(1):225–248
Florida R, Pedigo S (2020) How our cities can reopen after the COVID-19 pandemic. The Avenue: https://www.brookings.edu/blog/the-avenue/2020/03/24/how-our-cities-can-reopen-after-the-covid-19-pandemic/. Accessed March.
Florida R, Rodríguez-Pose A, Storper M (2021) Cities in a post-COVID world. Urban Stud 00420980211018072
Government of South Africa (2023) National Building Regulations and Building Standards Act 103 of 1977. Available at: National Building Regulations and Building Standards Act 103 of 1977 | South African Government (www.gov.za). Accessed on 9 April 2023
Habelsberger BE, Bhansing PV (2021) Art galleries in transformation: is COVID-19 driving digitisation?. In: Arts, vol 10, No 3. MDPI, p 48
Hedding DW, Greve M, Breetzke GD, Nel W, Van Vuuren BJ (2020) COVID-19 and the academe in South Africa: not business as usual. S Afr J Sci 116(7–8):1–3
Hoesli M, Malle R (2022) Commercial real estate prices and COVID-19. J Europ Real Estate Res 15(2):295–306
Iloms E, Ololade OO, Ogola HJ, Selvarajan R (2020) Investigating industrial effluent impact on municipal wastewater treatment plant in Vaal, South Africa. Int J Environ Res Public Health 17(3):1096
Iroham CO, Salubi DO, Okagbue HI, Oletubo AA (2022) Assessing void in Grade-A office buildings in Lagos metropolis. In: IOP conference series: earth and environmental science, vol 993, No 1. IOP Publishing, p 012003
Issaka RW, Bansah AK, Kuuder CJW (2022) Hotel managers' perspectives of Airbnb operations in Accra Metropolis, Ghana. Res Hospitality Manag 12(1):35–43
Keke OV, Okafor JI, Sado RO (2020) Influence of the novel coronavirus disease on Nigerian real estate sector and the way forward. Int J Real Estate Stud 14(S2):206–212
Kirk P (2019) Obsolete office buildings offer investors a Myriad of opportunities for adaptive re-use. Available at: Plenty of Alternative Uses Exist for Struggling Office Buildings (wealthmanagement.com). Accessed on 3 April 2023

Kotze N, Carow S (2019) Converting office blocks to affordable housing in the inner city of Johannesburg: social housing or a ruse for inclusive housing. In: Proceedings of the international scientific symposium new trends in geography, pp 203–213

Lambourne T (2022) Valuing sustainability in real estate: a case study of the United Arab Emirates. J Property Investment Finance 40(4):335–361

Lee A (2020) Wuhan novel coronavirus (COVID-19): why global control is challenging? Public Health 179:A1

Lukasiewicz M (2020) Johannesburg's financial globalization (c. 1880–1910) & South African financial archives. Social aims of finance, p 259

Madeddu M, Clifford B (2021) Housing quality, permitted development and the role of regulation after COVID-19. Town Plann Rev 92(1):41–48

Makinde OO (2020) J Human Earth Future

MAI SFF (2015) Market analysis for real estate. Appraisal J 83(1):77

Marutlulle NK (2021) A critical analysis of housing inadequacy in South Africa and its ramifications. Africa's Publ Serv Delivery Perform Rev 9(1):16

Mbunge E (2020) Effects of COVID-19 in South African health system and society: an explanatory study. Diabetes Metab Syndr 14(6):1809–1814

McKay T, Coumbias M, Kotze N (2021) Leisure cycling entrepreneurialism in Johannesburg, South Africa

Mkentane L (2023) Joburg property values not justified by the market, estate agents say. Available at: Joburg property values not justified by the market, estate agents say (businesslive.co.za). Accessed on 14 April 2023

Mukherjee S, Narang D (2022) Digital economy and work-from-home: the rise of home offices amidst the COVID-19 outbreak in India. J Knowl Econ 1–22

Naidu T (2020) The COVID-19 pandemic in South Africa. Psychol Trauma Theory Res Pract Policy 12(5):559

Navsaria PH, Nicol AJ, Parry CDH, Matzopoulos R, Maqungo S, Gaudin R (2021) The effect of lockdown on intentional and non-intentional injury during the COVID-19 pandemic in Cape Town, South Africa: a preliminary report. S Afr Med J 111(2):110–113

Nguse S, Wassenaar D (2021) Mental health and COVID-19 in South Africa. South Afr J Psychol 51(2):304–313

Obi SE, Yunusa T, Ezeogueri-Oyewole AN, Sekpe SS, Egwemi E, Isiaka AS (2020) The socio-economic impact of covid-19 on the economic activities of selected states in Nigeria. Indonesian J Soc Environ Issues (IJSEI) 1(2):39–47

Orekan A, Adebayo AJ (2020) Evaluation of flexibility in building design on commercial property value in Ilupeju, Odi-olowo LCDA Lagos state. Evaluation

Owolabi TJ, Ajibose KA (2019) Work-family conflict: coping strategies to optimise healthy living and wellbeing among career couples in Lagos metropolis

Pataki-Bittó F, Kapusy K (2021) Work environment transformation in the post COVID-19 based on work values of the future workforce. J Corp Real Estate 23(3):151–169

Profitable Venture Magazine (2021) Top 20 small business investment opportunities in Johannesburg. Available at: Top 20 small business investment opportunities in Johannesburg—ProfitableVenture South Africa. Accessed on 3 April 2023

Quirk J (2021) Making Johannesburg the epicentre. Camb Rev Int Aff 34(6):888–894

Rahman SME, Mele MA, Lee YT, Islam MZ (2021) Consumer preference, quality, and safety of organic and conventional fresh fruits, vegetables, and cereals. Foods 10(1):105

Ramagole DA, Janse van Rensburg DC, Pillay L, Viviers P, Zondi P, Patricios J (2020) Implications of COVID-19 for resumption of sport in South Africa: a South African sports medicine association (SASMA) position statement. South Afr J Sports Med 32(1):1–6

Rattanaprichavej N (2014) Enhancing the real estate development concept of highest and best use: a PEST analysis approach. In: Proceedings of the international symposium on business and management (ISBM), Nagoya, Japan

Remoy HT, van der Voordt DJM (2014) Adaptive reuse of office buildings into housing: opportunities and risks. Build Res Inf Int J Res Develop Demonstration 42(3):381–390
Research and Markets (2023) South Africa commercial real estate market—growth, trends, COVID-19 impact, and forecast (2023–2028). Available at: South Africa Commercial Real Estate Market—Growth, Trends, COVID-19 Impact, and Forecast (2023–2028) (researchandmarkets.com). Accessed on 9 April 2023
Rogerson CM, Rogerson JM (2020) COVID-19 tourism impacts in South Africa: government and industry responses. Geo J Tourism Geosites 31(3):1083–1091
Rymarzak M, Siemińska E, Sakierski K (2022) Reflecting sustainability in the analysis of highest and best use: evidence from polish municipalities. Real Estate Manag Valuation 30(4):103–115
Saladino V, Algeri D, Auriemma V (2020) The psychological and social impact of Covid-19: new perspectives of well-being. Front Psychol 2550
Scheba A, Turok IN (2020) Strengthening township economies in South Africa: the case for better regulation and policy innovation. In: Urban forum, vol 31, No 1. Springer, Netherlands, Dordrecht, pp 77–94
Schröder M, Bossert A, Kersting M, Aeffner S, Coetzee J, Timme M, Schlüter J (2021) COVID-19 in South Africa: outbreak despite interventions. Sci Rep 11(1):4956
Sekyere E, Bohler-Muller N, Hongoro C (2021) The impact of COVID-19 in South Africa. Available at: The Impact of COVID-19 in South Africa | Wilson Center. Accessed on 9 April 2023
Simpson T (2020) Adaptive reuse as a conduit for regenerative & human-centred design in Johannesburg
Spaull N, Van der Berg S (2020) Counting the cost: COVID-19 school closures in South Africa and its impact on children. South Afr J Child Educ 10(1):1–13
Sudharma IWP, Rahmadani DA (2020) Highest and best use analysis: case study of vacant land in Denpasar. Adv Glob Econ Bus J 1(2):1–8
Taiwo OJ (2022) Modelling the spatiotemporal patterns of urban sprawl in Ibadan metropolis between 1984 and 2013 in Nigeria. Model Earth Syst Environ 8(1):121–140
Todes A, Harrison P, Rubin M, Appelbaum A (2020) Thinking densification from and through Johannesburg. In: Densifying the city?. Edward Elgar Publishing, pp 17–39
Tokazhanov G, Tleuken A, Guney M, Turkyilmaz A, Karaca F (2020) How is COVID-19 experience transforming sustainability requirements of residential buildings? A review. Sustainability 12(20):8732
Ugwuonah C (2022) Department of architecture & planning, NED University of engineering & technology, city campus Maulana Din Muhammad Wafai Road, Karachi 32(2):1
Verma AK, Prakash S (2020) Impact of covid-19 on environment and society. J Glob Biosci 9(5):7352–7363
Walacik M, Renigier-Biłozor M, Chmielewska A, Janowski A (2020) Property sustainable value versus highest and best use analyzes. Sustain Dev 28(6):1755–1772
Walker R (2021) Covid-19 and the commercial rental market in SA. Available at: Commercial property sector struggling under impact of Covid-19, recession—survey (iol.co.za). Accessed on 9 April 2023
Wekeza SV, Sibanda M (2019) Factors influencing consumer purchase intentions of organically grown products in Shelly Centre, Port Shepstone, South Africa. Int J Environ Res Public Health 16(6):956
Wilke J, Mohr L, Tenforde AS, Edouard P, Fossati C, González-Gross M, Sánchez Ramírez C, Laiño F, Tan B, Pillay JD, Pigozzi F (2021) A pandemic within the pandemic? Physical activity levels substantially decreased in countries affected by COVID-19. Int J Environ Res Public Health 18(5):2235
Wilkinson S, Remoy H (2017) Adaptive reuse of Sydney offices and sustainability. Sustain Build
Yusuf R, Muhammad IH, Otunola AO, Kayode SJ (2021) Effect of accessibility on commercial property rental values performance in Oja Oba market Ilorin Metropolis. Path Sci 7(12):2001–2006
Zhang Y (2020) 20-century building adaptive-reuse: office buildings converted to apartments

Zumstein D, Oswald C, Brauer C (2021) Online retailer survey 2021: empirical findings on the e-commerce boom in Switzerland and Austria

Chapter 9
Neglected Third Estate: The Dilemma of Public Cemeteries in Osogbo Nigeria

Emmanuel Ademola Komolafe and Solomon Pelumi Akinbogun

Abstract As with other public facilities, cemeteries must be maintained on a regular basis to sustain their original state and to guarantee that they continue to fulfill their intended purpose. However, urban administration in Nigeria tends to place a low premium on burial spaces, as evidenced by the dilapidated physical conditions of public cemeteries. The purpose of this research is to assess the physical and environmental aspects, as well as management procedures, of two public cemeteries in Osogbo. Primary and secondary sources of data were used in this study: cemeteries and their surrounds were visited, and data on cemetery administration were received directly from the municipal officials. The questionnaire completed by (10) cemetery staffs was examined using the Relative Important Index (RII). The findings indicate that the two cemeteries are in a neglected state; are constrained by a shortage of experienced employees, a lack of contemporary equipment and management procedures; and are virtually always overgrown, isolating them from the urban environment. Additionally, the most significant obstacles impeding good preservation practices are a lack of a maintenance culture, a lack of current equipment, poor usage of the cemetery, a shortage of experienced staff, and sabotage by third parties. The recommendations include a complete makeover of these facilities and the establishment of a modern management team tasked with the objective of developing environmentally friendly, physically appealing, and motorable cemetery.

Keywords Cemetery · Maintenance culture · Neglect · Management · Planning

E. A. Komolafe (✉) · S. P. Akinbogun
Department of Estate Management, School of Environmental Technology, Federal University of Technology Akure, P.M.B 704, Akure, Ondo State, Nigeria
e-mail: komolafeea@futa.edu.ng

S. P. Akinbogun
Biomimicry and Nature-Based ICT Solution for Sustainable Construction Research Centre, Department of Built Environment, Faculty of Engineering, Built Environment and Information Technology, Walter Sisulu University, East London, South Africa

S. P. Akinbogun et al. (eds.), *Facility Management practices*,
https://doi.org/10.1007/978-3-031-65321-6_9

9.1 Introduction

Public cemeteries are part of the arrays of public facilities usually owned and maintained by government. Public facilities are those that are provided, owned, or supervised by the government at the federal, state, or local level (Iwarere and Lawal 2011). Cemeteries, like public facilities, can be owned by the federal, state, or local government. Cemeteries are a necessary component of a government's infrastructure. The national cemetery in Abuja is a classic example, as it contains the graves of slain national heroes. Cemeteries are intricate and fascinating locations with a plethora of meanings and purposes. They primarily provide culturally acceptable and sacred burial grounds for the living to memorialize their ancestors. As such, cemeteries hold cultural, personal, historical, social, and religious value for the communities in which they are located. Cemeteries that are densely vegetated can also provide city dwellers with the benefits of nature by serving as restorative, healthful green spaces in increasingly congested urban contexts. Graves are not merely places of burial that serve the public, they are green urban spaces with cultural and natural characteristics that might be integrated into the planning system for green infrastructure.

When referring to the 'sleeping place of the dead,' Omosulu and Ajegunmo (2013) referred to it as 'the third estate of man' and noted that town planners had lost sight of its inclusion in planning methods, particularly in poor nations. While death is a natural occurrence, many people are averse to discussing it. As many people are born daily, many also die, making it necessary to consider the allocation of land use for the disposal of the deceased, which, if not properly handled, can constitute a threat to man's health, psychological well-being, and environmental wellness. From the municipal perspective, burial is viewed as a critical and required public function mandated by law. The purpose of the statutory provision is to ensure that deceased humans are disposed of in a manner that does not pose a risk of epidemic breakouts. Obviously, the cemetery should be well-located, well-designed, well-maintained, and well-managed. The need to treat the dead with dignity is not only universally understood, but also transcends cultures and societies.

However, it is appalling that in Nigeria, the case is different. Oyebamiji (2018) observed that the state of cemeteries in Nigeria is very unkempt, deficiency in planning and lacking in maintenance. Douglas (2013) reported that public cemeteries are ancient and overstretched, that used graves have been 'unknowingly' re-used as bones have been pulled up while digging graves for burial, and that public cemeteries are overcrowded. Omosulu and Ajegunmo (2013) observed that in Owo town in Ondo State, there is high neglect of public cemeteries by officials in charge and in Nigeria as a whole. Also, the practice of burying the dead at homes has been discouraged by medical experts. This paper examines public cemetery management with the goal of providing a rational guide to policymakers in order to design a more effective maintenance practice approach for public cemeteries. It examines the physical state of cemeteries and the issues that impede successful cemetery administration.

Additionally, it considers how the physical shape of a cemetery may affect landscape management practices, public accessibility, and the cemetery's multifunctional purposes.

9.2 Literature Review

This section reviews the existing literature that is pertinent to the study, offering a thorough examination of the function and administration of cemeteries in urban environments. It investigates the potential of cemeteries as valuable green spaces in densely populated cities and analyses a variety of factors that influence the planning and management of these spaces.

9.2.1 Cemeteries as a Resource for Green Space

Over half of the world's population now lives in cities (United Nations 2018), and as cities get denser, open green spaces are created (Haaland and van den Bosch 2015). As cities become more densely populated, cemeteries may offer much-needed urban green space in undeveloped neighborhoods and cities (Harnick 2010). Landscapes of urban cemeteries, particularly those with horticultural resources, have gained prominence (Charap and Barnett 2016). Civic cemeteries, created to assure a decent and sanitary burial, were historically located close but not inside towns (Rugg 2000), but cities have expanded beyond their boundaries. This stance may drive cemeteries to evolve from being purely burial locations to being available to the public for leisure reasons (Skar et al. 2018). Cemeteries have developed into multi-use landscapes as a consequence of changing attitudes in response to urban expansion constraints and the limited availability of open space (Harvey 2006). Cemeteries have traditionally been primarily spiritual settings, owing to their primary role of burying the departed. While cemeteries are sacred burial places, their value varies throughout time and across cultures (Rugg 2000). Due to cemeteries' unique and holy significance within civilizations, their landscapes are typically preserved as a gesture of respect for the buried, under the concept that cemeteries will offer everlasting resting places for the dead. The historical and cultural significance of cemeteries also helps ensure their long-term existence.

9.2.2 Planning and Management of Cemeteries

Cemeteries are public sites designed by urban planners. When planning a cemetery, it is critical to remember that, like other land uses, cemeteries are not governed by the same set of rules as other land uses due to their inherent differences. As a result, numerous jurisdictions throughout the world have enacted legislation governing the layout and construction of their separate cemeteries (Ndung'u 2015). As a result, The London Planning Advisory Committee (1997) specified that cemetery planning and administration must be directed by defined cultural, religious, humanitarian, social, economic, and environmental objectives and must address such strategic concerns as burial space selection, cost of disposing of the deceased by whatever methods, closeness to sites of remembrance, open space for green chains and natural habitats, historic elements, and biodiversity. Additionally, Auckland Council (2013) emphasized the need of incorporating sustainability into cemetery planning and administration by ensuring that burial grounds are sustainable in all aspects of well-being. These ideas include social, which refers to a social system's capacity to function at a high degree of well-being and harmony; cultural, which exhibits social equality and the safeguarding of cultural values; environmental, which requires providing huge expanses of green space; and economic, which entails contributing to the local economy and developing the local skill base. Indeed, sound design and administration may help cemeteries balance competing needs for space, environmental preservation, and economic activity (Ndung'u 2015).

McDougall and Vine (2005) established parameters for evaluating cemetery planning and management. These parameters include the provision of planning and management policy guidelines, the development and maintenance of available infrastructure within cemeteries, the establishment of staff and safety policies, the provision of public utilities, the preservation of heritage value and character, landscaping and environmental management, and stakeholder and public participation. Additionally, McDougall and Vines asserted that design and management of cemeteries, whether public or private, are not particularly difficult to execute.

In Nigeria, cemetery planning has gotten insufficient attention, as repeated by Omosulu and Ajegunmo (2013), who assert that urban planners have lost sight of planning for the dead, particularly in the majority of poor countries. However, Aneirobi et al. (2016) argued that the 1999 constitution of the Federal government of Nigeria mandates local governments to offer public cemetery services. Aneirobi et al. (2016) ended the research by stating that, at the present, the local government has made no effort to prepare for cemeteries in the studied area. Additionally, in Osun State, local government authorities have lost sight of and made little effort to plan for cemeteries, as grave grounds within the city are reserved for strangers or accident victims. As a result, deceased remains are buried in unapproved locations across the area.

9.2.3 Public Cemeteries in Nigeria

Numerous studies such as Akinbogun et al. (2020): Akinbogun and Jones (2018), Akinbogun (2017), had been conducted to demonstrate the poor state of public cemeteries in Nigeria and the dangers associated with interment at home. Other studies have documented Nigeria's failure to adequately house the dead and have urged the government to intervene immediately. Olajide et al. (2013) examined the issue of burying the deceased in Nigeria and its prospects. Using chi square and frequency state analysis, the study investigated present cemeteries, burial patterns, and factors influencing people' cemetery selection. The investigation discovered that housing for the deceased is insufficient and in certain cases, terribly untidy. Additionally, that people prefer to bury their deceased in their homes rather than in cemeteries. This issue is particularly prevalent in Nigeria, owing to the people's cultural and religious beliefs. The paper went on to urge that public–private partnerships be formed to preserve established public cemeteries and to encourage investors to capitalize on the graveyard shortage. Douglas (2013) conducted a physical inspection of all five public cemeteries in Rivers state. The study discovered that public cemeteries in Rivers state are underdeveloped in terms of infrastructure and operational maintenance. The study cautions that these unmaintained cemeteries may represent a health risk to state citizens. Onwuanyi et al. (2017) used a frequency table and facility assessment method to investigate three public cemeteries in Benin City. The investigation discovered that the public cemetery, which is an integral element of the metropolitan landscape, has been neglected by local government officials.

9.3 The Study Area

Osogbo is Osun State's capital city, located in southwestern Nigeria. It is situated between the latitudes 7°46′N and 7.767°N and the longitudes 4°34′E and 4.567°E and has an area of 47 km^2. Additionally, Osogbo city serves as the headoffice of both the Olorunda Local Government Area (Osogbo North). and the Osogbo Local Government Area (Osogbo South). Osogbo has a population of approximately 316,000 NPCs, according to the 2006 census (2006). She is home to a variety of local and modern public and private institutions and industries. Osogbo is primarily occupied by Yoruba people who work as farmers, artisans, and office employees. Two public cemeteries are found in Osogbo, one in the powerline area and one in the Owode area.

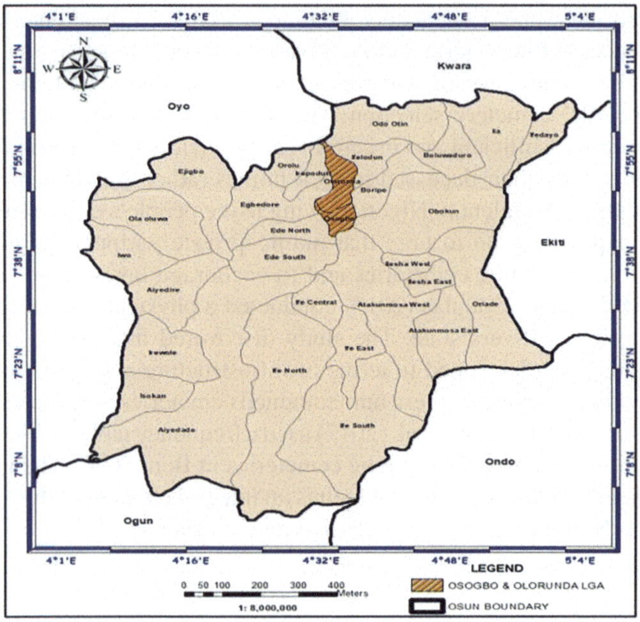

Map of Osun showing Osogbo. *Source* Adapted from Atoyebi et al. (2020)

9.4 Research Methodology

The term "data" refers to the information that is gathered from an object that must be processed. The transformed data is transformed into information that is useful for action. As the primary data set for this journal, a well structured questionnaire was used to collect data on management issues through interactions with officials from the Osogbo Local Government Environmental Department who are responsible for administering the cemeteries. Physical investigations of the graves were done/ conducted to ascertain their condition. Respondents were chosen using purposeful sampling techniques. Ten cemetery managers were questioned in all. Additionally, data were gathered through observations, direct interviews, and the review of published and unpublished research. To explain and emphasize the results of the hypotheses investigated, a frequency table, and simple statistical analysis were used. The relative importance index was used to determine the factors impeding good management practice. The relative importance index (RII) was used using the following equation to estimate the importance assigned to the variables affecting the maintenance practice. Additionally, the relative relevance indicators' levels were determined.

$$\text{RII} = \sum (\text{X1} * \text{Y1} + \text{X2} * \text{Y2} + \text{X3} * \text{Y3} + \ldots \text{Xn} * \text{Yn})/(\text{A} * \text{N})$$

where X represents the response category weights, Y the frequency of responses, A the highest weight, and N the total number of responses.

9.5 Discussion and Findings

This section presents the results and findings of the research. It includes a detailed analysis of the condition of cemeteries and the factors affecting effective maintenance practices. Say something here.

9.5.1 Condition of the Cemeteries and State of Cemetery Management

Physical observations were made of each of the two cemeteries. The results of the physical inspections are presented below (Table 9.1).

The conditions of Powerline Cemetery and Owode Cemetery are compared in the table based on four factors: Environment, Structure, Facilities, and Security. Both cemeteries are severely neglected. The tombstones of Powerline Cemetery are scarcely visible due to the encroachment, while Owode Cemetery is also plagued by overgrowth, an unpaved, rough road, and untidy environs. Structurally, neither cemetery is demarcated or fenced, which presents security risks and potential encroachment issues. The absence of guiding indications, paved paths, internal roads, and essential utilities such as electricity and water supply in both facilities impedes maintenance and accessibility. Both cemeteries are devoid of security personnel, which enhances the likelihood of unauthorised activities and vandalism. The table

Table 9.1 State of the cemeteries

Elements of inspection	Powerline cemetery	Owode cemetery
Environment	Overgrown, with, tombstones barely visible	Unpaved, rough road, untidy surroundings, overgrown
Structure	Not fenced and no demarcation	Not fenced and no demarcation
Facilities	There are no guiding indications and there are no paved paths/internal roads. There is no evidence of electricity or water supply	There are no guiding indications and there are no paved paths/internal roads. There is no evidence of electricity or water supply
Security	No security seen	No security seen

Source Field Survey (2022)

Table 9.2 Staffing of the cemeteries in Osogbo

Role of staff	Powerline cemetery	Owode cemetery
Supervisor	1	1
Clerk	2	1
Diggers	3	2
Total	6	4

Source Field Survey (2022)

Table 9.3 Equipment used by cemetery grounds men

Type of equipment	Purpose
Pick axes	Excavation
Shovels	Excavation/infilling
Wheel barrows	Removal of excavated soil
Cutlasses	Bush clearing

Source Field Survey (2022)

emphasises the severe dereliction and inadequacies in both cemeteries, suggesting the necessity of comprehensive management improvements, such as regular maintenance, infrastructural enhancements, and security measures, to ensure orderliness and respect.

Cemeteries are still administered using antiquated methods that are labour expensive, technologically archaic, and lacking in innovative concepts. Indeed, cemetery management policy is as dormant as cemeteries themselves. Overgrown plants, dilapidated fences, and the absence of basic amenities all speak to the story of neglect. According to officials, the now-overgrown cemetery have been cleansed only once in the last two years in response to public concerns. Cemeteries provide the impression that they are locations where the deceased are laid to rest and then forgotten. This may explain why they are not supplied by public utilities (electricity, water) and why, except for the modest huts at their entrances, there are no amenities on site that provide shelter of any kind for tourists or even the cemetery's few workers. The present staffing level, as seen in Tables 9.2 and 9.3, reflects the cemeteries' operating capacity (human resources) and capabilities (equipment) to service a city population of over 316,000 and to manage the cemeteries' huge area, which needs sizeable standing grounds maintenance team equipped with contemporary equipment.

9.5.2 Factors Militating Effective Maintenance Practice

This section explores the factors that impede the efficacy of cemetery maintenance practices. It is imperative to comprehend these obstacles in order to devise strategies that will enhance the maintenance and administration of cemeteries. The Relative

Table 9.4 The relative importance index of factors militating effective maintenance practice

Factors	RII	Rank
The lack of a culture of maintenance	0.92	1st
Lack of modern equipment	0.88	2nd
Improper use of the cemetery	0.84	3rd
Lack of skilled workers	0.75	4th
Sabotage by third party	0.70	5th
Insufficient funds	0.68	6th
Political problems	0.62	7th
The lack of institutional and training facilities	0.59	8th
Attitude of workers	0.49	9th
Security aspects	0.38	10th

Source Field Survey (2022)

Importance Index (RII) of a variety of factors is illustrated in the subsequent table, which emphasises their severity and ranking.

From Table 9.4, with a relative importance index of 0.92, lack of a culture of maintenance ranks 1st as factor militating effective management practice. This clears shows that lack of maintenance cross across all government facilities. In addition, with a relative importance index of 0.88, 0.84, 0.75, 0.70 and 0.68, lack of modern equipment, improper use of cemetery, lack of skilled workers, sabotage by third party and insufficient funds all place 2nd, 3rd, 4th, 5th and 6th respectively on the relative importance list. While Political problems, lack of institutional and training facilities, altitude of workers and security aspects came 7th, 8th, 9th, and 10th respectively.

A comprehensive framework for the maintenance of public cemeteries is necessary to address these factors. Policy interventions should encompass the development of a maintenance culture through community engagement and training, as well as the modernization of equipment and the establishment of appropriate use guidelines. It is also imperative to enhance security measures to prevent sabotage, ensure adequate funding through alternative mechanisms such as public–private partnerships, and develop a competent workforce. The maintenance and administration of public cemeteries can be substantially enhanced by the implementation of these strategies.

9.6 Conclusion

Assuming that public cemeteries in Nigeria are routinely neglected, as demonstrated by their poor look and published comments, this paper studied the state and management status of two public cemetery in Osogbo. The findings reveal that all cemeteries are neglected and poorly maintained; that a comprehensive makeover is essential;

and that new, environmentally sensitive cemeteries with contemporary design and management standards are required to complement the urban landscape. Additionally, the lack of maintenance culture was noted as a major problem in the cemeteries. As a result, it is proposed that the following be done: First, the public cemeteries in Osogbo be rebuilt to blend with the urban environment rather than present a rural image. Second, cemetery administration would gain from the use of contemporary administrative methods and technology, resulting in greater resource management, record keeping, managerial capability, and accountability.

References

Adeboye O (2016) Home burials, church graveyards and public cemeteries: transformation in Ibadan mortuary practice, 1853–1960

Akinbogun SP (2015) Modelling residential tenants' choices with a grave as a negative externality (Doctoral dissertation, Heriot-Watt University)

Akinbogun SP (2017) An analysis of NIMBY solution and tenants' property right to a home with a grave in the private rented sector. J Contemp Res Built Environ 1(1)

Akinbogun SP, Jones CA (2018) Externalities and planning failure in the housing market of an African City. Int J Hous Market Anal 11(3)

Akinbogun SP, Aigbavboa C, Gumbo T, Thwala W (2020) Modelling the Socio-economic implications of sustainability issues in the housing market, published by Springer. https://www.springer.com/gp/book/9783030489533

Anierobi CM, Nwokolo CN, Nwalusi DM, Umeh AO, Okeke SO (2016) Urban land use planning and facilities: a reflection on cemetery in Enugu Metropolis, Nigeria. J Trop Environ 13(1)

Atoyebi OS, Oyeniyi SO (2020) Assessment of provision, adequacy and spatial distribution of green infrastructure in Oshogbo. Int J Sci Acad Res 1:6–12

Armstrong S (2012) Collegia and mortuary archeology: identifying the non-elite of the Roman empire. J Manitoba Anthropol Student's Assoc 30:1–16

Charap J, Barnett D (2016) When cemeteries are also public gardens. Public Gardens 31(4)

Douglas KE (2013) Current state of public cemeteries in Rivers state, Nigeria. Afr J Med 3(14):58–61

Haaland C, van Den Bosch CK (2015) Challenges and strategies for urban green-space planning in cities undergoing densification: a review. Urban For Urban Greening 14(4):760–771

Harnick P (2010) The future of the European chemical industry. KPMG

Harvey T (2006) Sacred spaces, common places: the cemetery in the contemporary American city. Geogr Rev 96(2):295–312

Iroham CO, Okagbue HI, Ekanem IF, Peter NJ, Samuel OW, Nto SE, Isiaka S, Adedotun AF (2024) Three attributes determining land values in three selected housing estates in Uyo, Nigeria. Front Sustain Cities 6:1384945

Iwarere HT, Lawal KO (2011) Performance measures of maintenance of public facilities in Nigeria. Res J Bus Manage 5(1):16–25

McDough and Vine (2005) Management plan guidelines for South Australian cemeteries. https://cdn.environment.sa.gov.au/environment/docs/management_plan_guidelines_for_sa_cemeteries_2005.pdf

Ndung'u JK (2015) A comparative analysis of planning and management of cemeteries in Nairobi, University of Nairo and Naad Libiua. Published thesis

Olajide SE, Alabi OT, Akinlabi T (2013) Problem and prospect of housing the dead in Nigeria. Int J Eng Inventions 2(1)

Omosulu S, Ajegunmo G (2013) The neglect of cemetery as a unit of planning in Owo, OndoState, Extract from conference proceedings, Faculty of Environmental Studies, Rufus Giwa Polytechnic, Owo, Ondo State, Nigeria, pp 448–462

Onwuanyi N, Ndinwa C, Chima P (2017) Public cemeteries of Benin City: examining a neglected dimension of Urban Nigeria. Niger J Environ Sci Technol 1(2):367–378

Oyebamiji IO (2018) Achieving sustainable final resting place for corpse disposal through effective planning and landscaping, unpublished master's thesis. Federal University of Technology Akure

Quinton JM, Duinker PN, Gallant KA, Steenberg JW, Charles JD (2019) To tree or not to tree: user and management perspectives of cemetery trees. Urban For Urban Greening 43:126385

Rugg J (2000) Defining the place of burial; what makes a cemetery. University of York, UK. Mortality 5(3). ISSN 1357-6275

Skar M, Nordh H, Swesen (2018) Green urban cemeteries: more than just parks. J Urbanism Int Res Placemaking Urban Sustain 11(3):1–21

United Nations (2018) Revision of world urbanization prospects. Retrieved from. https://www.un.org/development/desa/publications/2018-revision-of-world-urbanization-prospects.html

Chapter 10
Analysis of the Causes of Rapid Post-construction Breakdown of Plumbing Facilities in Ghana

Adu Gyamfi Timothy, Emmanuel Yaw Frempong, Jnr, Pim-Wusu Mark, and Akorli Solomon King

Abstract The quality of plumbing materials in Kumasi the capital of Ashanti Region in Ghana is questionable. Plumbing fittings deteriorate quickly after installation. The study's purpose is to assess the causes of poor plumbing work in the built environment industry in Ghana. The research utilised stratified and purposive sampling methods to solicit respondents' responses. Data were collected from 10 plumbing teachers, 40 plumbers, 15 suppliers of plumbing materials, and 20 clients through a self-completed questionnaire with a five-point Likert scale. The data were analysed with a one-way analysis of variance (ANOVA) at a 0.05 level of significance. A post hoc test was used to compare the significant differences that existed among the views of the stakeholders. Findings reveal that Ghana's standard board does not check the quality of plumbing materials before they arrive in the country. There is no association to monitor the inflow of inferior plumbing fittings. The study found that most plumbers are not equipped with knowledge of the basic plumbing principles of vent pipe termination requirements. Most plumbers do not know the reasons for constructing inspection chambers in drainage systems and fixing traps to the sanitary appliances. The study contributes to the literature on the causes of poor plumbing in the built environment industry in Ghana.

Keywords Quality plumbing material · Plumbing principles · Plumbers · Kumasi · Ghana

A. G. Timothy (✉) · E. Y. Frempong, Jnr · P.-W. Mark
Faculty of Technical Education, Department of Construction, Akenteng Appiah-Menka University of Skills Training and Entrepreneurial Development, Kumasi, Ghana
e-mail: agttimo78@gmail.com

E. Y. Frempong, Jnr · A. S. King
Faculty of Built Environment, Department of Building Technology, Koforidua Technical University, Koforidua, Ghana

P.-W. Mark
Department of Building Technology, Accra Technical University, Accra, Ghana

S. P. Akinbogun et al. (eds.), *Facility Management practices*,
https://doi.org/10.1007/978-3-031-65321-6_10

10.1 Introduction

While a building's circulatory system is its property, a person's body is the domain of their plumbing system. Your life will be thrown wildly and made miserable by any flaw in the system. A well-designed plumbing system should be installed since it is important. In order to get oxygen, nutrients, and other things to the body's numerous cells, blood must be moved throughout the body through the cardiovascular system. Humans live because of those veins and arteries. Similar pipes throughout the home provide water to various rooms, assisting people in maintaining their cleanliness and hydration. Blockages can lead to serious issues in the body as well as the home. Even though a plumbing problem isn't nearly as dangerous as a heart attack, there are some startling parallels between pipes and heart health (Futuramics Clean Water Centre 2017). A home's health and safety, in particular, can be greatly impacted by its plumbing system. Every homeowner can learn the value of a well-working plumbing system after experiencing one or becoming involved in one (Residencestyle 2019). The spread of diseases, which is inevitable with a subpar plumbing system because wastewater is replete with bacteria, parasites, and other pathogens, is one of the potential safety and health effects of a subpar plumbing system. A broken plumbing system may cause the environment to become more contaminated with disease-causing organisms and hazardous substances, as well as contaminate groundwater, drinking water, and food. This could have an impact not only on the immediate area but also on nearby areas (Residencestyle 2019). Leaking pipes can persist for a long time after they first started to leak, making them one of the most important problems that property managers need to address right away. In these situations, leaking pipes have the potential to seriously harm the building's structural integrity, escalating the scope of necessary repairs from minor plumbing issues to significant construction problems (Jeff 2018). Leaking pipes under the sink and on the walls are most likely to cause mould, which is the biggest issue. Mould can begin to grow in as little as 24–48 h if there is enough moisture (Alley 2010). An entire building may suffer damage if pipes inside leak for a long time. Flooding, warped and stained walls, invasive black mould, and water quality issues are just a few of the damages that leaking pipes can create (Best Plumbing Service 2016). Hugo (2007) discovered that dampness causes damage and that 70% of construction issues are directly or indirectly related to moisture as an unseen harm that lessens the strength and rigidity of fabric.

According to Wong et al. (2008), the consequences of a poor drainage system are not just complaints from building occupants about bad smells; they also include the spread of bacteria and viruses that can result in illness or, in the worst case, fatalities. The severe acute respiratory syndrome (SARS) virus was found to have been spread by a droplet from virus-rich excreta that returned to residents' apartments via sewage and drainage systems where there were strong upward air flows and inadequate plumbing due to non-functional water seals in traps, according to the findings of the WHO investigation into the Amoy Gardens outbreak in 2003 (WHO 2003). Because viral aerosols were present in drainage stacks during COVID-19 outbreaks in high-rise buildings, Guo et al. (2022) hypothesised that faecal aerosol inhalation was a likely method of transmission. According to the World Plumbing Council (WPC), the likelihood of a major health catastrophe is constantly rising in a world where large numbers of people routinely travel between countries and continents. Poor plumbing practises were the cause of the SARS outbreak in Hong Kong; therefore, it is crucial that all countries maintain high standards for sanitation and plumbing in order to prevent a repeat of this situation. Governments could promote the industry's role in promoting public health by ensuring that the department in charge of or sponsoring it is one that deals with health, as this would help raise awareness of the subject (Burgon 2008).

The widespread issues with plumbing technology in Kumasi, the capital city of the Ashanti Region in Ghana, are extremely concerning. Some of the issues with the plumbing fixtures in both public and private buildings are due to the quick degrading of the plumbing appliances soon after handover. Poorly designed above- and below-ground drainage systems in public and private residences lead to obstructions in access to clearing blockages in high-rise structures and uncomfortability in accessing plumbing facilities. There is no literature on this area, and after years of witnessing widespread issues with plumbing technology in the Kumasi Ashanti Region of Ghana, it was therefore thought important to look into the reasons for the quick post-construction collapse of plumbing infrastructure in Ghana, hence this study. The goal of the study is to ascertain how materials affect plumbing installations and plumbers' familiarity with fundamental plumbing concepts. When questions are given in research work as a guide to coming to a very definitive conclusion, the deeper reflection comes into focus. The following are the research questions that will direct the study's attention:

1. What are the stakeholders' opinions about the impact of materials on plumbing installations?
2. Do the stakeholders' opinions on how materials affect plumbing installations differ?
3. How do the stakeholders feel about building plumbers' capacity based on fundamental plumbing principles?
4. Do the stakeholder's opinions on building plumbers' capacity based on fundamental plumbing principles differ?

10.2 Quality and Durability of Plumbing Materials

The pipes, couplings, fittings, and other plumbing system components should all meet recognised quality and performance criteria and be strong enough to provide reliable service for a long time. On the product, indicators of standard compliance are frequently marked (Health Aspects of Plumbing 2006). Bell (2023) opined that using good plumbing materials is important to maintain the integrity of the system and to ensure convenience and safety. It is also important from an aesthetic standpoint (Lennwy 2019; Bell 2023). Lennwy (2019) advocated that choosing the right plumbing supplies is all about protecting customers biggest investment and enhancing the value and functionality of a building. However, using cheap, low-quality faucets or fixtures, for instance, may result in spending more money to replace them before long (Bell 2023). Again, buying inferior quality usually means that the material is far more likely to deteriorate rapidly and fail quickly (Lennwy 2019).

Any pipelines, valves, taps, or other fittings used to supply drinking water or remove wastewater must not have any dangerous materials present in amounts greater than those allowed to leach into the water. The fittings, including the pipes, valves, taps, and others, must be strong enough to withstand predicted internal pressures while being able to carry water at the designated pressure within the specified environment. If they are to be buried, they also need to be able to bear outside pressure. It's also important to take into account the effects of environmental variables including temperature, humidity, pressure, corrosion, and levels of bacteria (Health Aspects of Plumbing 2006). Quacco (2022) finds out that the quality and standards of building materials are compromised due to a lack of supervision and monitoring by the Ghana Standard Authority.

"All structures must be constructed to accommodate plumbing systems," says Hall (1993). However, draughtsmen in Ghana design the majority of privately held buildings without the involvement of architects. As a result, plumbing designs are no longer up to code. Sobieski (2022) asserts that the design and implementation of numerous complex systems are necessary for the construction of a new structure. A plumbing engineer should be a member of the project team to guarantee that these systems are correctly planned and implemented. Any remodelling job that requires a considerable amount of plumbing work must have a well-planned diagram. For instance, a plumbing layout drawing allows one to plan out the system in advance in a kitchen or bathroom, which will help ensure the process goes as smoothly as possible with the help of a professional plumber, who typically meticulously plans a plumbing job to avoid surprises during the project (Sole 2022a). A plumbing layout shows how the water and waste removal systems will work. Drains, vent pipes, valves, and fixtures like toilets and sinks are frequently included in it (Sole 2022a). Sasaki and Kawashima (2010) state that assessments of the risk of shear failure as a result of the main rod being severed have been made. According to Sasaki and Kawashima (2010), when longitudinal bars are reduced in the portion of the structure where the largest moment may occur, it will eventually damage the structure's shear strength and result in future shear cracks. While doing what should have been done before the wall and floor are built, plumbers drill holes through walls and floors to pass 75- to 150-mm pipe. These activities run the risk of damaging sensitive building components. Al-sibahy and Edwards (2021) contend that a significant weakening of walls results from a deep vertical or horizontal chase. The steel frame of the building, which is a component of the structure's base, is where moisture poses the biggest threat to structural integrity through seepage. Despite the fact that structural steel is meant to survive for numerous lifetimes, if it is not adequately protected, the metal will corrode. And when that occurs, it enlarges. If the steel corrosion is not fixed, it can lead to a variety of structural damages, including cracks in both interior and exterior walls, broken door arch bars, falling masonry, and displacement of door and window frames. The corrosion also causes concrete slabs to separate from the foundation (Best Plumbing Service 2016).

10.3 Building Capacity of Plumbers on Basic Plumbing Principles

Building capacity is the process of enhancing the knowledge, talents, procedures, and resources that enable organisations and communities to endure, adapt, and prosper in a world that is undergoing rapid change. Transformation that is created and sustained over time from within is a crucial component of capacity-building; this type of transformation extends beyond completing tasks to changing mindsets and attitudes (United Nations Academic Impact 2022). The achievement of human development and security depends on capacity building, says the Global Development Research Centre (GDRC) (2020). It is a procedure that includes networking, activities with multiplying effects, value-added education, and the training of trainers. Building the capacity of individuals, businesses, industries, and communities aimed to improve the knowledge, communication, and managerial abilities required to address the emerging issues in its programmes and themes and to promote information dissemination among local communities, policymakers, academics, researchers, individuals, and other institutions. The World Plumbing Council (WPC) (2006) defines plumbing as work related to the provision of water supply pipe work, including, without limitation, firefighting, all types of space heating, sanitary appliances, discharge pipe work, and drainage. If it weren't for the convenience that plumbing systems gave us, our lives would have been considerably more demanding and challenging. Plumbing gives us control over water, protects us from epidemic diseases, stops water waste, and offers comfort and beauty (Benjamin 2019). For a home to remain comfortable, the plumbing system is essential. It would be difficult to take a shower, wash clothes, use the toilet, wash dishes, irrigate the lawn, or use the dishwasher if the plumbing system was not running properly (Lavergne's 2020). When it fails, the plumbing system in homes and businesses almost invariably provides us with reliable service. Then, when anything goes wrong, we realise exactly how much we depend on a dependable plumbing system (Plumbing Dynamics 2022). Plumbers have a massive amount of work to do; as a result, they require intelligence to solve complex equations, expertise and responsibility to install various systems, knowledge of how and why various systems function, the ability and initiative to solve problems, a thirst for learning about new technology, a creative mind, good judgement, and most importantly, professionalism and honesty. Only when their capacities are established in the fundamentals of plumbing can plumbers effectively display these traits. Plumbers with formal and informal schooling should brush up on their knowledge of fundamental plumbing concepts. Three components make up a home's plumbing system:

vent pipes, sewer lines, drainpipes, and traps. It's crucial to understand these components. These three components are the plumbing system's heart and soul, and since they can malfunction, one needs to be aware of them (Ezinemark 2012).

Drain-waste-vent (DWV) pipes smoothly transport waste and water out of the house without gurgling or odours. Behind the water, there must be an air channel. In order to offer that passage and to remove odours from the house, vent pipes run from the drainpipes up through the roof (Sole 2022b; Treloar 1996). To ensure that drain water flows smoothly, vent piping supplies air pressure. Each fixture has a vent pipe that extends through the roof and is connected to the drain lines. They are covered to deter animals and rain while allowing air to enter (Copeland 2021). According to the Health Aspect of Plumbing (2006), the vertical ventilation stack pipe both prevents offensive odours from entering the building and serves as an outlet to the outside air. Additionally, it makes sure that no vacuum or pressure builds up in the drainage system that could suction or blast the liquid seal from the fixture traps. In some places, putting the stack pipe against the building's outside wall is the most popular way to install it in single-family homes. The vent pipe's terminal is extended upward at least 300 mm from the roof eaves by passing through or around the eaves (of a sloping roof).

This height should be extended to at least 600 mm if the roof is accessible and flat. The stack pipe needs to be far away from any windows. Where it is impracticable to carry any internal stack above roof level, it may be taken through the building's outside wall and carried upward to terminate at least 3 m (10 feet) horizontally from a window or other opening of the building or of any neighbouring building (Health Aspect of Plumbing 2006; Blower 2000). As most vent pipe insulation is subpar, as shown in Plate 10.1, it is imperative that the building construction sector spread knowledge regarding vent pipe termination.

The plumbing connection's primary component, the plumbing trap, is intended to drain water into the stack. The trap also protects against clogging and serves as a barrier to keep sewage odours out of the room (MasterProf 2022; Masterman

Plate 10.1 Poor termination of vent Pipe at window opening

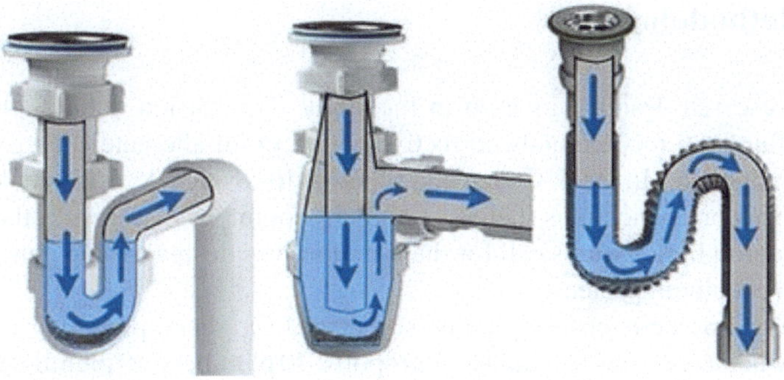

Plate 10.2 Different traps and their water seal

and Boyce 2000). A trap is a unique component made to link plumbing fixtures to wastewater disposal systems in order to form an airtight plug between the sewer and the room, preventing odours and bacteria from entering the room (MasterProf 2022; Blower 2000). The Amoy Gardens incident was characterised by a lack of maintenance that allowed floor drains serving bathroom facilities to get dry, creating a pathway for contaminated air to enter habitable space (John 2009). There are fittings for larger trap swings, couplings to use between unaligned pipes, and flexible extension couplings for modifications that need to be made vertically or horizontally (Stickley 2022). Plate 10.2 shows different traps and their water seals.

Drainage systems are a sanitary and clean way to remove waste from bathrooms, toilets, and dirty water in homes (Civil Engineering Bible 2022). A drainage system consists of pipes that transport sewage, rainwater, or other liquid waste to a site of disposal, such as a septic tank or the sewer system. Its major goals are to gather and eliminate wastewater and prevent sewer gases from entering the structure (Copeland 2021). Although drainage systems are beneficial for public health, there is considerable potential for pipe blockages. Manholes and the inspection chamber system are necessary to reduce the chance of blockage. Any underground drainage system must include inspection chambers and manholes. They serve as an entry point, enabling maintenance personnel to examine, clean, and clear drains of any obstructions. The obstruction in the subsurface pipes can be reached directly through a manhole. Even though the inspection chambers are modest, they only permit cameras and other devices to examine and test the drainage and water flow via the pipes (Civil Engineering Bible 2022; Easy Merchants 2022). Every time a horizontal drain changes its direction or grade, when two drains converge, or when the length of a drain exceeds 30 m (100 feet), an access chamber should be built (Health Aspects of Plumbing 2006). When there is a change in pipe size, when a straight run of piping is longer than 22 m, when a drain changes direction by more than 30°, or when a branch pipe joins a main drainage run inside the drainage system, inspection chambers and manholes are employed (Easy Merchants 2022).

10.4 Methodology

A survey research design was used in the study. This design outlines the current state of affairs and recommends corrective action or an alternate course of action. (Alright, unpublish.) In 1995, Cohen and Manion discovered A survey looks into the circumstances or connections that exist and are common. As a result, the ongoing issues with plumbing fixtures call for the use of survey design to help generate ideas for the current circumstances.

The study's participants are plumbers, material suppliers, plumbing instructors, building owners, or clients in Kumasi Metropolis. 40 plumbers, 10 plumbing teachers, 20 clients, and 15 plumbing suppliers make up the study's sample size. Asamoah and Duodo (2007) asserted that a sizable sample size—typically at least 30—is necessary for normal distribution. Purposive and stratified random sampling were the sample techniques employed for the investigation. The target population was included in the study through purposive sampling to ensure knowledgeable responses in the subject area, whereas stratified sampling is employed when representation from all the subgroups is necessary and the measurement of interest differs between the subgroups. 85 respondents, separated into four strata, were given a self-completed survey to complete. The survey was broken down into three sections: section A gathered demographic information; section B gathered information on the impact of materials on plumbing installations; and section C gathered data on plumbers' familiarity with fundamental plumbing concepts. The study was conducted using a five-point scale: strongly disagree (1), disagree (2), not sure (3), agree (4), and highly agree (5). (Anderson 1985; Casley and Kumar 1988; Downie 1967). Utilising the statistical software for the social sciences, version 26.0, the data were examined item by item (SPSS 26.0). A priori, a significance level of 0.05 was established. For each of the four categories of respondents, the mean rating for each item was calculated. Then, in order to ascertain whether respondents agreed with the statement on the feature of the programme represented by the questionnaire item, each computed mean rating was compared with the theoretical mean rating (assuming a normal distribution of responses) of 2.5. To ascertain whether there were any statistically significant variations in the stakeholders' opinions of the study, an item-by-item test analysis for paired means of the stakeholders' responses was conducted. A one-way analysis of variance (ANOVA) with $p = 0.05$ was considered statistically significant with regard to the perspectives of the stakeholders. The turkey post hoc test was used to do pairwise comparisons of the means to identify differences in the evaluations of instructors, plumbers, customers, and suppliers where significant differences were identified. ANOVA makes the assumption that the data are normally distributed and that the variances are homogeneous. When there is obvious heterogeneity in variances, Kinnear and Grey (1992) strongly advocate previous proof of the homogeneity assumption and recommend the use of nonparametric tests. The variables used to measure the study were tested to determine their reliability. The results of the reliability test had a Cronbach's alpha value of 8.95, which indicates the attainment of data dependability.

10.5 Result and Discussion

Concerning the stakeholders' opinions about the impact of materials on plumbing installations, findings shows that each of the four subgroups for each item's mean value and associated resulting mean were calculated as part of the study. To determine whether or not the objective had been accomplished, the computed mean was contrasted with the 2.5 theoretical mean value. The median value was higher than 2.5 (Table 10.1), indicating that all the variables are influencing how materials are used in plumbing installations. The item with the highest final mean value was Ghana Standard Board doesn't examine quality plumbing materials before they enter the country (3.86), followed by no association to monitor the influx of inferior fittings (3.62). Using high-quality plumbing materials to help customers feel at ease has a mean value of 3.33, while using quality plumbing materials to make plumbing work easier has a mean value of 3.28. The least significant resultant mean (2.63) is related to the fact that some plumbers have trouble identifying high-quality plumbing materials. Despite the high level of the final mean value for each of the eight (8) items used to gauge the impact of materials on plumbing installations, it was thought appropriate to use one-way analysis of variance (ANOVA) at the 0.05 level of confidence to see if there were any statistically significant differences in the respondents' ratings. Table 10.1 displays the item-by-item ANOVA outcome (the last two columns). A turkey post hoc test analysis with significance levels of 0.05 was utilised to analyse the stakeholder viewpoints in order to identify which ones substantially differ from one another. The findings indicate that seven of the eight items were not statistically significant. There is not a major variation in opinion between all those involved. The fact that the Ghana Standard Board does not inspect high-quality plumbing supplies before they enter the nation was generally agreed upon by respondents, showing that this is true. This assertion is in line with Quacco's (2022) report that the quality and standard of building materials are compromised because of a lack of supervision and monitoring by the Ghana Standard Board. Again, the study discovered that the use of quality plumbing materials makes clients feel comfortable, which agrees with Lennwy's (2019) advocate that selecting the right plumbing supplies goes a long way to protecting customers biggest investment and enhancing the value and functionality of a building. Also, the findings reveal that using quality plumbing materials makes plumbing work easier, which agrees with Bell's (2023) opinion that using good plumbing materials is important to maintain the integrity of the system and to ensure convenience and safety. In addition, the study finds the following: plumbing materials on the market are not of the best quality as required; using quality plumbing materials prolongs the life of a building; and most customers are unwilling to buy quality plumbing materials. However, there is no association to monitor the inflow of subpar plumbing fittings. In determining how materials may affect plumbing systems, these elements are crucial. However, one item was statistically significant based on the responses of the respondents, suggesting that stakeholders had different opinions on the element. This is because plumbers find it challenging to discover high-quality plumbing materials.

Table 10.1 Stakeholders view about the effect of materials on plumbing installations

Statement	Plumbers N = 40		Teachers N = 10		Client N = 20		Suppliers N = 15		Resultant mean	Ranking	*F*. value	*P*. value
	Mean	SD	Mean	SD	Mean	SD	Mean	SD				
The use of quality plumbing materials make client feel comfortable	3.50	0.51	3.00	0.00	3.50	0.53	3.30	0.48	3.33	3rd	1.89	0.14
Using quality plumbing materials make plumbing work easier	3.80	0.51	3.00	0.00	3.10	0.88	3.20	0.63	3.28	4th	1.61	0.19
Using quality plumbing materials prolong life span of building	3.24	0.44	3.00	0.00	2.80	0.42	3.50	0.53	3.14	6th	1.82	0.07
Plumbing materials at the market are not of the best quality as required	3.48	0.51	3.60	0.56	2.60	0.52	3.00	0.00	3.17	5th	1.93	0.16
Most clients are not willing to buy quality plumbing materials	3.56	0.51	2.60	0.89	2.50	0.85	3.00	0.00	2.75	7th	1.98	0.07
Plumbers find it difficult to identify good plumbing materials	3.44	0.51	3.00	0.00	1.80	1.03	1.60	0.99	2.63	8th	2.79	0.04

(continued)

Table 10.1 (continued)

Statement	Plumbers N = 40		Teachers N = 10		Client N = 20		Suppliers N = 15		Resultant mean	Ranking	F. value	P. value
	Mean	SD	Mean	SD	Mean	SD	Mean	SD				
No association to monitor the inflow of inferior plumbing fittings	3.68	0.48	4.00	0.48	3.40	0.52	3.40	0.51	3.62	2nd	1.32	0.28
Ghana standard board don't check quality plumbing materials before it arrives in the country	3.80	0.41	4.00	0.00	3.84	0.42	3.81	0.63	3.86	1st	1.83	0.18

Regarding how the stakeholders feel about building plumbers' capacity based on fundamental plumbing principles, each item's mean value across its four subgroups was calculated, as was the item's resulting mean rating. The calculated mean was contrasted with the 2.5 theoretical mean value. Any mean value below 2.5 denotes disagreement with the factors, while a mean value above 2.5 denotes agreement with the factors. Table 10.2 demonstrates a variable with a higher resulting mean value of 3.86, which is associated with a lack of plumbers' awareness of vent pipe termination; a mean value of 3.84 is associated with a lack of plumbers' knowledge of using inspection chambers in drainage systems; a lack of plumbers' knowledge of the purpose of fixing traps to appliances has a mean value of 3.64; and a lack of plumbers' knowledge of determining the height of overhead tanks has a mean value of 3.59.

The outcomes of the item-by-item ANOVA are shown in Table 10.2. (The last two columns) A turkey post hoc test analysis at 0.05 significant levels was utilised to compare the stakeholder perspectives in order to identify which ones substantially differ from one another. The results show that six of the eight elements were not statistically significant in relation to the responses of the respondents, indicating that there is no disagreement among the stakeholders. As a result, respondents rated a lack of plumbers' knowledge of vent pipe termination, a lack of plumbers' knowledge of using inspection chambers in drainage, a lack of plumbers' knowledge of the purpose of fixing traps to appliances, and a lack of plumbers' knowledge of fundamental plumbing principles. On the other hand, two of the elements—a lack of knowledge of how to determine the height of overhead tanks and plumbers' lack of knowledge of the uses of elbow and bend—were statistically significant with regard to the responses of the respondents, suggesting a difference in the views of stakeholders regarding those two elements.

10.5.1 Implication of the Study

The study implies that the Ghana Standard Board should intensify its supervision and monitoring of quality plumbing materials that are brought to Ghana. This document can also be used as the basis for developing policies that will assist in setting up a committee to monitor the quality of plumbing materials in Kumasi, Ashanti Region of Ghana. This study will help the Ghana Plumbers Association embark on a sensitization programme on the concept of plumbing principles and the quality of plumbing materials in the Ghana plumbing industry. Also, socially, the study seeks to promote responsible plumbing installation that will avoid practises that promote the egress of sewer gas into our environment, which further leads to health problems.

Table 10.2 Stakeholder's views on building capacity of plumbers on basic plumbing principles

Statement	Plumbers N = 40		Teachers N = 10		Client N = 20		Suppliers N = 15		Resultant mean	Ranking	*F*. value	*P*. value
	Mean	SD	Mean	SD	Mean	SD	Mean	SD				
Lack of plumbers' awareness of basic plumbing principles	3.48	0.51	3.50	1.10	3.30	1.21	3.27	1.03	3.39	5th	1.64	0.08
Lack of plumbers' knowledge of the purpose of fixing trap to appliance	3.83	1.13	3.70	1.21	3.60	0.71	3.59	1.05	3.69	3rd	1.90	0.06
Provision of many elbows in supply pipe networks	2.94	1.34	3.00	1.55	2.89	0.48	3.15	0.97	3.00	8th	1.98	0.07
Lack of plumbers' knowledge of the uses of elbow and bend	3.25	1.08	3.20	1.04	3.32	0.53	3.16	0.00	3.23	7th	3.92	0.04
Lack of knowledge of determine height of overhead tanks	3.56	0.51	3.60	0.67	3.49	0.84	3.72	0.00	3.59	4th	4.82	0.02
Lack of plumbers' ability to interpretate basic plumbing designs	3.34	0.54	3.10	0.56	3.40	0.74	3.23	0.82	3.27	6th	1.73	0.18

(continued)

Table 10.2 (continued)

Statement	Plumbers N = 40		Teachers N = 10		Client N = 20		Suppliers N = 15		Resultant mean	Ranking	*F*. value	*P*. value
	Mean	SD	Mean	SD	Mean	SD	Mean	SD				
Lack of plumbers' awareness of vent pipe termination	3.97	1.02	3.88	0.92	3.82	0.48	3.78	0.84	3.86	1st	1.89	0.08
Lack of plumbers' knowledge of using inspection chambers in drainage	3.89	0.41	4.10	0.70	3.74	0.421	3.62	0.63	3.84	2nd	1.98	0.09

10.6 Conclusion

The majority of stakeholders in public or private organisations have expressed serious concerns about the quick degeneration of plumbing fittings following the transfer of plumbing work. Because the materials used for plumbing work quickly degrade after being used, this has led to the subpar plumbing work witnessed in Kumasi Metropolis. The Ghana Standard Board doesn't inspect high-quality plumbing materials before they enter the country; using high-quality plumbing materials helps customers feel at ease, and using quality plumbing materials makes plumbing work easier, according to the study's findings about the effects of materials on plumbing installation. The study also found that the areas that needed plumbers' capacity to be built on fundamental plumbing principles included a lack of awareness of vent pipe termination, a lack of knowledge of using inspection chambers in drainage systems, a lack of knowledge of the purpose of fixing traps to appliances, and a lack of knowledge of calculating the height of overhead tanks. The study made the recommendation that Ghana needs a strong institutional framework or policy governing the practise of plumbing technology to help control the importation of subpar plumbing products. To help decide who should practise plumbing in the country, Ghana should establish standards for sanitation and plumbing, and the government organisations tasked with monitoring the quality of plumbing supplies imported into the nation should step up their oversight to prevent the importation of subpar plumbing supplies. The study is restricted to Kumasi Metropolis, and additional research might be carried out in other Ghanaian metropolises as well as in the Ashanti area and elsewhere.

References

Anderson LW (1985) Likert scales. In: International encyclopaedia of education, vol 5. Pergamon, Oxford, UK, pp 3082–3084

Alley A (2010) Bad plumbing and the effect of leaking pipes. Alley Article

Al-Sibahy A, Edwards R (2021) Effect of chases with renovation techniques on the load carrying capacity of MasonryWalls. Infrastructures 6:160. https://doi.org/10.3390/infrastructures6110160

Asamoah-Gyimah K, Duodu F (2007) Introduction to research methods in education. The Institute for Educational Development and Extension University of Education Winneba

Bell Plumbing and Heating (2023) Benefits of using good plumbing supplies. https://bellplumbing.com/blog/2013/04/benefits-of-using-good-plumbing-supplies/?sfw=pass1685310234. Access on 28 May 2023

Benjamin F (2019) Reasons why plumbing is important for health, safety and water sustainability. https://www.benjaminfranklinplumbing.com/rosenberg/about-us/blog/2019/january/reasons-why-plumbing-is-important-for-health-saf/

Best Plumbing Service (2016) Leaking pipes: causes, problems, and solutions. https://www.bestplumbercincinnati.com/blog/2016/july/leaking-pipes-causes-problems-and-solutions/. Access on 21 Nov 2022

Blower GJ (2000) Plumbing mechanical service book 2, 3rd edn. Logman limited, London

Burgon DR (2008) The role of the plumbing industry in global heal. World Plumbing Council, Scotland

Casley D, Kumar K (1988) The collection, analysis, and use of monitoring and evaluation data. The John Hopkins University Press, New York

Civilengineeringbible (2022) Inspection chambers in drainage system. https://civilengineeringbible.com/article.php?i=195

Downie NM (1967) Fundamentals of measurements: techniques and practices, 2nd edn. Oxford University Press, London

Easy merchants (2022) What is an inspection chamber? Our guide. https://www.easymerchant.co.uk/blog/inspection-chamber-guide/

Ezinemark (2012) Understanding the plumbing system. https://newsghana.com.gh/understanding-the-plumbing-system/?amp. Access on 21 Nov 2022

Futuramics Clean Water Center (2017) How is plumbing like your cardiovascular system? https://omahawater.com/plumbing-like-cardiovascular-system/

Guo Y, Li X, Luby S, Jiang G (2022) Vertical outbreak of COVID-19 in high-rise buildings: the role of sewer stacks and prevention measures. Current Opinion in Environ Sci Health 29:100379. https://doi.org/10.1016/j.coesh.2022.10037

Hall FE (1993) Plumbing technology, 2nd edn. Logman limited, London

Health Aspects of Plumbing (2006) by the World Plumbing Council and the World Health Organisation, Geneva

Hugo SLC (2007) Building physics heat, air and moisture: fundamentals and engineering methods. Ernest and John Wiley company, London

Jeff A (2018) How leaking pipes can cause structural issues. https://remedial.com.au/blogs/how-leaking-pipes-can-cause-structural-issues/. Access on 21 Nov 2022

John S (2009) Air pressure and drainage cross-contamination: Ezine Articles

Kinnear PR, Gray CD (1992) SPSS/PC+ made simple. Lawrence Erlbaum, Hove

Lavergne's Plumbing heating (2020) The significance of plumbing in day-to-day life. https://lavergneplumbing.com/the-significance-of-plumbing-in-day-to-day-life/

Lennwy Crosstown Plumbing (2019) The benefit of using high-quality plumbing supplies. https://www.crosstownplumbing.com/the-benefits-of-using-high-quality-plumbing-supplies/. Access on 28 May 2023

Masterman AH, Boyce (2000) Plumbing mechanical service book 1,2&3. Stanley Thornes publishes limited, Chelten Ham

MasterProf (2022) Plumbing trap: types, how to assemble, which one to choose. https://masterprof.net/en/academy/articles/plumbing-trap-types-how-to-assemble-which-one-to-choose/

MT Copeland (2021) What is a drainage system in plumbing? https://mtcopeland.com/blog/what-is-a-drainage-system-in-plumbing/

Plumbing Dynamics (2022) The three main reasons most of plumbing systems fails. https://plumbingdynamicsdallas.com/3-main-reasons-most-plumbing-systems-fail/

Quacco EB (2022) Poor quality of building materials: Ghana Standards Authority urged to strengthen its surveillance. Joy Business new. https://web.facebook.com/watch/?v=540687701192688. Access on 28 May 2023

Residencestyle (2019) The importance of good plumbing for your home. https://www.residencestyle.com/the-importance-of-good-plumbing-for-your-home/

Sasaki T, Kawashima K (2010) Scale effect on the shear strength of reinforced concrete columns with termination of main reinforcements. J Struct Mech Earthquake Eng JSCE 66(1):187–205

Sobieski Plumbing Heating and Air Conditions (2022) 5 reasons why you need a plumbing engineer when constructing a new building. https://www.sobieskiinc.com/blog/5-reasons-why-you-need-plumbing-engineer-when-constructing-new-building/

Sole C (2022a) How to draw a plumbing plan for your next remodeling project. https://www.bhg.com/home-improvement/plumbing/how-to-draw-a-plumbing-plan/

Sole C (2022b) Everything you need to know about venting for plumbing work. https://www.bhg.com/home-improvement/plumbing/drain-venting/

Stickley A (2022) How to use flexible drain parts. https://www.thespruce.com/about-flexible-drain-parts-2718771

The global development research center (GDRC) (2020) Capacity building policy. https://www.gdrc.org/about/cb-policy.html
Treloar RD (1996) Plumbing encyclopedia, 2nd edn. Black well science limited, London
United Nation Academic Impact (2022) Capacity building. https://www.un.org/en/academic-impact/capacity-building
WHO (World Health Organization) (2003) Final report: Amoy gardens. WHO Environmental Investigation. www.iapmo.org/common/pdf/ISS-Rome/SARS_Amoy_Gardens_HK.pdf
Wong ESW, Chan DWT, Jones P, Law LKC (2008) Assessment of Air Pressure inside a Drainage Stack. Proceedings of the International Multi-Conference of Engineers and Computer Scientists, Hong Kong, 19–21 March 2008, 1534–1539

Chapter 11
Assessment of the Key Factors Responsible for the Shortage of Skilled Artisans in the Nigerian Construction Industry

Akinbogun Oluwaseun Tope

Abstract The growing shortage of skilled artisan in the construction industry calls for concern. This paper examines the causes of the issue with a view to provide a direction for policy makers in the industry. The research design was basically survey. It involved a 2 rounds Delphi survey among 25 selected expert in the construction industry. Data were collected by simple random sampling from 370 stakeholders in the construction industry by means of structured questionnaire. It applied descriptive statistics for the analysis of data. The essential elements are assessed and ranked according to their average ratings. The findings indicate that the primary cause of the scarcity of skilled artisans in the Nigerian construction sector is the insufficient investment in staff and manpower development, as evidenced by a mean score of 4.46. This is closely followed by a lack of interest in skill acquisition, with a mean score of 4.42.

Keywords Artisans · Skill · Shortage · Causes · Construction · Industry

11.1 Introduction

The building industry and its associated activities are widely recognized as a primary driver of economic progress. According to Ofori (2012), the construction industry constituted a significant portion, ranging from 20 to 50%, of a nation's overall wealth. Additionally, it accounted for approximately 10–14% of household expenditure. Ernst and Sarabia (2015) believe that the construction business exhibits distinct characteristics in terms of its functions and links with other sectors, resulting in a notable multiplier impact compared to other industries. Specifically, the construction industry predominantly comprises of locally operating, average-sized contractors

A. O. Tope (✉)
Faculty of Engineering and Built Environment, Department of Operation Management, University of Johannesburg, Johannesburg, South Africa
e-mail: seunakinbogun@yahoo.com; otakinbogun@futa.edu.ng

S. P. Akinbogun et al. (eds.), *Facility Management practices*,
https://doi.org/10.1007/978-3-031-65321-6_11

who primarily focus on residential projects. The construction services and engineering services business play a vital part in the economic advancement and growth of the nation. The construction business, on a worldwide scale, currently employs a significant portion of the global labor force, accounting for approximately 7% of total employment. According to Mwanaumo (2014). A crucial role is played by the construction industry in the economic growth and employment creation of a nation (Ernst and Sarabia 2015). Noteably, the sector is a crucial aspect of the economy which plays the role of social and economic advancement of any given nation (Theisohn and Lopes 2011; Okoye 2016). Ofori (2012) noted that the importance of construction to any economy can be attributed to increased connection with the other economy. It is however noted in a survey conducted in South Africa that people that are gainfully employed by the construction industry are well above 1.4, be it permanent or contract (South African Construction 2015). For countries that are really industrialised, about 20% of the GDP could be responsible for by the construction industry and it could also be responsible for about 12% of its workforce employment creation. 3.2% of the GDP is contributed by the sector in Nigeria (Dantata 2007; Construction Overview 2011). The construction industry is the major support of the Nigeria economy (Olowa 2018). This is made possible because virtually all other sector of the economy depends solely on the construction sector to carry out their functions. The construction sector has been essential in its linkage with other sector of the country's economy thereby having multiplier effect on them especially the real estate as well as the building materials (Muyiwa 2003). The study of human capacity development was applied by Carson (2005), Kululanga (2005), CIDB (2002), which says that the word capacity transmits to the capability, abilities, awareness, knowledge abilities, principles, connections, conducts, enthusiasms, resources and situations that allows a person, establishments, organisations and structures to perform their responsibilities effectively, effectually and inventively for their growth objectives to be attained. In essence, capacity development comprises several methods and technics that look for ways of expanding performance (Kululanga 2015). The capacity building of construction industry is seen as the cautious and accomplished method which enhances the influence the industry in achieving nation-wide, construction demand, the promotion of general, social and economic advancement of the industry there by leading the enactment, effectiveness, and improved value to stakeholders CIDB (2002). This undertaking discourses the role and the efforts of those who contributed to the of the processes beginning from commencement of the project to the handing over of the project and the maintenance, which comprises of the community and reserved sector clients, professionals in built environment, constructors, producers of materials and contractors, training delivery establishments, governing bodies and enquiry institutes (Kululanga 2012: 86–100). According to Groot and Molen (2000), the availability and quality of skilled workforce play a critical role in determining the performance of the construction industry. Nevertheless, Yamoah (2014: 139) has identified signs of scarcities and diminished proficiency among construction labourers in Nigeria. Many studies have conducted an examination of human capacity development across many fields However, there is limited knowledge exists regarding the key factors responsible for shortage of artisans in the construction industry that influence the enhancement of artisans' capacity in developing nations such as Nigeria. The study design a comprehensive descriptive analysis of skill shortage among artisans within the construction sector in Nigeria. This is achieved through the use of a prominent methodology, known as delphi survey (DS), which is employed for the purposes of data collecting, and analysis.

11.2 The Shortage of Skilled Artisans in the Construction Industry

According to Ofori (2001) and Okuntade (2014), skill shortage is a major problem confronting the Nigeria construction industry and threatening the economy of the nation. The shortage of skilled labour affects all areas of construction especially in the areas of time, cost and the quality of work (Medugu et al. 2011). This view was corroborated by Dantong et al. (2011) who noted the shortage in skilled craftsmen is not the shortage of labour but rather it is the lack of trained productive labour that are available for specific jobs. According to Ajagbe et al. (2012), the shortage of skilled labours are qualitative and the quantitative. The qualitative phase is, a situation whereby the skilled labours are not occupying the vacancies or that there are a few of them while, the later which is the quantitative implies that employers may complain of shortage of workers when, the available workers do not possess the required skills. According to these studies, the factors responsible for the shortage of skilled craftsmen are; rapid change in technology, meagre remuneration of skilled craftsmen, lack of interest to young potentially skilled workers, aging of skilled workforce in the industry, inadequate training of skilled craftsmen and lack of motivation to skilled craftmen.

Rapid Change in Technology: It is observed that the construction industry all experiencing a rapid change in technology. The increase rate of rapid change in technology faced by the construction industry as viewed by Dubem et al. (2012) is due to increase in the use of technology. This has made it essential for organizations to meet the changing situation in the industry. The construction industry around the world has since been responding to the change in order to meet the demand of the industry but the majority of the construction industry in Nigeria are yet to adapt to the change thereby bringing a constraints on the Nigeria construction workforce and the industry at large (Okuntade 2014).

Meagre Remuneration of Skilled Craftsmen: The poor compensation of the construction craftsmen is a crucial problem facing the industry as skilled craftsmen are not attracted to the industry as there are no regulations guiding the minimum wage of construction workers. In the Nigeria construction industry, the construction craftsmen are being paid different wages across the country, the wage that is being paid to craftsmen in a state differs from that which is being paid to another in another state which has prompted construction workers to seek for other career or migrate to neighbouring countries where they will be better paid (Fagbenle 2004).

Lack of Interest to Young, Potentially Skilled Workers: The construction industry is not appealing to the younger generations in Nigeria thereby bringing about poor image to the industry. The work done by crafts men is seen to be done by less intellectual craftsmen. The poor image and career path has since been seen as a discouraging factor to youths and not seeing the construction industry as a career path for them to ply. Nigerian youth are no longer interested in acquiring skills especially construction related skills compared to that of developed nations like the United States of America (USA) as indicated that the youths demanding to acquire apprenticeship skills are more than the available training centres (Awe 2006).

Elderly Skilled Workforce in the Industry: In recent time, the average age of trained artisans and craftsmen in the Nigeria construction industry is between the ages of 45–55 years which is one of the major challenges facing the sector as there are little or no skilled youths to replace the aging workforce. As said by Dantong et al. (2011), if the trend is not given adequate attention, there will be little or no skilled artisans and craftsmen in the nearest future.

Inadequate Organization Training and Retraining of Skilled Craftsmen: For any nation to have a sustained economic growth, her human capital needs to be given adequate attention as it is an asset any organization, hence, the need for training for individual capacity development (Bilau et al. 2015). In the Nigerian construction industry, attention is usually not being paid to workers but majorly on the financial gain of the industry which makes it one of the major problems or cause of the shortage of skilled craftsmen in the industry (Dantong 2007). The lack of craftsmen training in any given organization brings about problems as there will be an increase in the competencies and the effectiveness of the workforce, therefore without the training of the workers, the problem mentioned will keep manifesting in the organization except a training policy is put in place by the organization (Onuka et al. 2012). It is therefore suggested that for there to be an end to the shortage of skilled artisans in the sector, training and development should be seen as proper tool for the updating of workforce development.

Skilled Craftsmen are Not Motivated: Motivation is an art of encouraging an individual to work optimally (Solomon et al. 2012). It is eminent that motivation propels an individual to put his or her best to work for an employer but in the case on the Nigerian construction industry, skilled workers are not being motivated to work for improved productivity which has always resulted in high staff turnover (Bilau et al. 2015). It was noted by Fagbenle (2004) that motivation of skilled staff can be acquired through a lot of method but economic reward must be the most considered. Ugheru (2006), contemplates that welfare package, incentives, job security amongst others are forms of considerations to aid motivation. Oni (2020) noted that there is need for significant improvement with respect to integration, individual participation, funding, recruitments and policy interventions. He is of the opinion that collaboration among key stakeholders and adopting best practices is important for alleviating the workforce challenge in the study area. Offei-Nyako et al. (2014) studied factors responsible for the scarcity of skilled artisanal in Ghana Construction Industry. Authors used relative important indices (RII) to investigate the importance of each factors relative to others. The factors in descending order of RII are: irregular and low remuneration, low motivation, the lack of interest by the youth to take up trades like masonship and carpentry as professions, varying working conditions and technological advancements. Thus, the study suggested promoting the welfare of its artisanal by the Association of Ghanaian Construction Artisans (AGCA). While these studies provide a good a background on the identification of the factors responsible for skill shortages among artisans, empirical evidence dissecting the impact of the effect is currently lacking in Nigeria.

11.3 Research Method

The utilization of the Delphi technique in this research facilitated the attainment of a consensus regarding the existing practices employed by the Nigerian construction industry for enhancing the skillset of craftsmen, as well as the evaluation of experts' perspectives on their effectiveness. According to the research conducted by Sackman (1974) and Linstone and Turoff (1975), Delphi is a communication strategy that is carefully constructed and intended to facilitate organized and participatory predictions. This method relies on the input and expertise of a panel of experts. Furthermore, it was employed in the identification of the primary causes and sub-factors that contribute to the scarcity of skilled craftsmen and their impact on the enhancement of artisans' ability in the building sector of Nigeria. The approach is founded on the notion of utilizing predictions made by a well selected panel of experts.

11.3.1 Selection of Expert and Data Collection

This study utilized the Delphi technique as a means to create the study model during its initial phase. The Delphi method often comprises multiple rounds, typically two to three, namely Round 1, Round 2, and Round 3. In this particular investigation, it was deemed essential to do two rounds of analysis subsequent to achieving a consensus during the second round. In Round 1, experts are solicited for their opinions on the matter under investigation in an open-ended manner. They are asked to rate the statements raised in a questionnaire according to their preferences on the subject. In Round 2, the panelists were requested to re-evaluate the rankings based on the results from Round 1. The determination of the final findings of the Delphi survey relies on the calculation of either the mean and the median of the final round. The identities of the panelists were undisclosed to one another, with just their perspectives being shared among the panel for the purpose of mutual awareness and decision-making. The methodology employed in this study involves the administration of a questionnaire to gather data.

11.3.2 Delphi. Process and Selection of Criteria

The initial round of the Delphi procedure involved the participation of 20 panels, while 17 experts were subsequently selected to continue their involvement in the second and third rounds of the study. During the expert selection process, emphasis was given on the experts' knowledge in the development of artisans' capacity, their high level of theoretical understanding, and their practical experience in the specific field of study. They were seen as being important as they were expected to have had a prior knowledge of artisans' capacity development. The panels were chosen to have some level of ideas in the field of capacity development, especially those of artisans.

Throughout the entirety of the Delphi survey, a questionnaire was constructed using the insights derived from the expert opinions obtained during the Delphi study. However, the initial phase of the survey consisted of questions derived from a condensed overview of the extensive literature analysis. This review identified a specific set of traits that are deemed crucial for achieving capacity development among craftsmen in the building industry of Nigeria. Moreover, the literature research encompassed the identification of many reasons contributing to the scarcity of proficient craftsmen within the building sector of Nigeria. Additionally, the examination of educational and training influences, along with other pertinent aspects, on the enhancement of artisans' competence was undertaken. The initial phase of the Delphi study employed a combination of open-ended and closed-ended questions. These questions were carefully examined and the outcomes served as the foundation for the subsequent rounds of the Delphi study. To assess the level of unanimity achieved by the experts, a descriptive statistics was used. Also, in order to reduce redundancy, responses gotten from the questionnaire were analysed using the content analysis methodology (Rubin et al.; Aigbavboa 2014; Somia 2019). The second round of the discussion served the purpose of allowing experts to thoroughly examine and provide feedback on the various factors that contribute to the development of artisans' skills in the Nigerian construction industry. Additionally, it provided a platform for experts to address pertinent issues related to the capacity development of artisans in this industry, as initially proposed by the participants in the first round. The second phase of the Delphi study employed closed-ended questions to examine participants' opinions regarding their agreement or disagreement with the identified traits that contribute to the development of artisans' capacity in the Nigerian building sector. The findings derived from the second round of the Delphi survey revealed a high level of consensus among the experts in their responses. The reason for the Delphi research technique was to attain a consensus among the experts. A result of each rounds of the Delphi was sent to the experts. They all got to know that an agreement was reached when all the participants are in 100% agreement. Also Aigbavboa

(2014) informed that when a higher percentage of the experts are all in agreement, therefore, there is a common consent and a common consent is acceptable in this study. When 60% of the expert agreed on a each of the questions, a common consent was said to be attained. Each of the questions were analysed individually in order to determine if consensus was attained or not based on the scale provided for each of the questions. Following the last round of the Delphi which was the last, a consensus was reached on all the factors responsible for the shortage of skilled artisan in the Nigerian construction industry.

11.3.3 Delphi Survey Round 3 Data Collection

The sample frame was all the building construction professionals who are corporate members of their respective professional bodies as at 2nd January 2023. They include; builders, quantity surveyors, architects, project managers, structural and civil engineers, estate surveyors and valuers, as well as craftsmen and artisan in the Nigeria compendium of artisans in Lagos and Akure. The professionals in construction as well as craftsmen are considered part of the sample frame to be considered in the factors that can bring about the artisans' capacity development in the Nigeria construction industry because they work hand in hand with the artisans as artisans cannot work in isolation of the construction professionals and craftsmen. They serve as trainers and mentors to the artisans in construction. The list was obtained from the states branch offices which are also available online at the website of the selected professionals. Due to the substantial size of this demographic, the respondents were selected using a probability sampling technique, specifically by sampling from a list of individuals within the population.

An examination of expert opinions regarding the causes of the shortage of skilled artisans in the Nigerian construction industry reveals that out of the thirty (30) subfactors identified, five (5) were found to have a very high impact (VHI) rating, ranging from 9.00 to 10.00. Additionally, twenty-four (24) factors were determined to have a high impact (HI) rating, ranging from 7.00 to 8.99, while one (1) item was classified as having a moderate impact (MI) rating, ranging from 5.00 to 6.99. The comprehensive variable was determined to have exerted influence on the causes contributing to the scarcity of skilled artisans in the building industry of Nigeria. The IQD score demonstrated a high level of agreement for the thirty (30) items, as all of them received scores ranging from 0.00 to 1.00. In terms of mean ranking, Table 11.1 shows that the primary factor contributing to the shortage of trained artisans in the Nigeria building industry is the absence of well-structured incentives, which holds the top position. Conversely, the inability to join trade unions is placed 30th among the variables responsible for this shortage.

Table 11.1 Key factors responsible for the shortage of skilled artisans

Causes of skilled artisans shortage	Median (M)	Mean ($\overline{x}$)	Standard deviation (σx)	Interquartile deviation (IQD)	Mean scores ranking (R)
Aging of skilled workforce in the industry					
Fewer workers to replace the aging workforce	8	7.47	2.09	0.00	20
Gradual extinction of artisans	8	7.63	1.16	0.50	19
Low influx of youths into the industry	9	8.68	0.89	1.00	4
Aged workforce	9	8.74	1.68	0.00	2
Lack of interest in technical skill acquisition	9	8.63	0.60	1.00	5
Poor remuneration of skilled artisans					
Absence of regulation guiding minimum wage	8	8.16	1.07	0.50	7
Migration of construction workforce	8	7.42	2.19	0.50	21
Workers are prompted to pursue other career	8	8.00	1.33	0.00	12
Inability to join trade union	6	6.11	1.52	0.50	30
Absence of workers' full loyalty	7	7.11	1.52	0.50	26
Lack of motivation of skilled craftsmen					
Lack of well-structured incentives	9	8.79	1.23	0.00	1
Lack of motivation for improved productivity	8	8.16	0.50	0.00	7
High staff turnover	8	8.16	0.96	0.50	7
Lack of an adequate wage policy	8	8.16	1.34	1.50	7
Lack of an adequate welfare package	8	8.16	1.07	0.50	7
Lack of appeal to young potential skilled workers					
Poor image associated with construction labour	8	7.78	1.90	0.75	16
Less intelligent craftsmen	7	6.72	1.84	0.75	29
Lack of interest in skill acquisition	7	7.33	0.97	0.75	23
Low demand for apprenticeship	8	6.83	2.85	0.75	28

(continued)

Table 11.1 (continued)

Causes of skilled artisans shortage	Median (M)	Mean ($\bar{x}$)	Standard deviation (σx)	Interquartile deviation (IQD)	Mean scores ranking (R)
Rework, cost and time overrun	7	7.28	1.27	1.00	24
Lack of organization training and retraining of skilled craftsmen					
Focus is on financial gains	9	8.72	1.23	0.00	3
Inadequate discussion on workforce improvement	7	7.39	0.78	0.00	22
Inadequate training policy	8	8.28	1.07	1.00	6
Absence of craftsmen training	8	7.78	1.26	0.75	16
Lack of staff and manpower development	8	7.89	1.13	0.00	14
Rapid change in technology					
Ever increasing sophistication	7	7.00	1.68	0.75	27
Globalization in the industry	7	7.17	1.79	1.00	25
Lack of response to sporadic technology	8	7.89	0.90	0.00	14
Credible consultants and contractors	7	7.67	1.50	1.75	18
Qualified and competent craftsmen	8	8.00	0.84	0.75	12

Source Delphi survey (2023) M = Median; x = Mean; σx = Standard Deviation (SD); IQD = Interquartile Deviation; R = Ranking

11.4 Discussion and Findings

The descriptive study of the important factors contributing to the scarcity of skilled craftsmen in the Nigerian construction sector (CSAS) is presented in Table 11.2. In addition to providing the rank of each variable, it shows the mean score and standard deviation for each variable. The essential elements are assessed and ranked according to their average ratings. Findings indicate that the primary cause of the scarcity of skilled artisans in the Nigerian construction sector is the insufficient investment in staff and manpower development, as evidenced by a mean score of 4.46. This is closely followed by a lack of interest in skill acquisition, with a mean score of 4.42, and an inadequate welfare package, with a mean score of 4.41. Some other important factors include the absence of craftsmen training (mean = 4.41), the presence of qualified and competent craftsmen (mean = 4.40), and low demand for apprenticeship (mean = 4.37). On the other hand, some of the factors with the lowest

Table 11.2 Descriptive analysis of key factors responsible for the shortage of skilled artisans in the Nigerian construction industry (CSAS)

Key factors	Mean	Std. deviation	Rank
Lack of staff and manpower development	4.46	0.674	1
Lack of interest in skill acquisition	4.42	0.684	2
Lack of an adequate welfare package	4.41	0.671	3
Absence of craftsmen training	4.41	0.657	4
Qualified and competent craftsmen	4.40	0.844	5
Less intelligent craftsmen	4.39	0.655	6
Inadequate discussion of workforce improvement	4.37	0.659	7
Low demand for apprenticeship	4.37	0.680	8
Globalization in the industry	4.36	0.774	9
Rework cost and time overrun	4.36	0.662	9
Lack of interest in technical skill acquisition	4.36	0.700	9
Lack of well-structure incentives	4.35	0.667	12
Focus is on financial gains	4.34	0.698	13
Lack of motivation for improved productivity	4.33	0.702	14
Aging workforce	4.31	0.766	15
High staff turnover	4.30	0.655	16
Lack of response to sporadic technologies	4.30	0.761	16
Inadequate training policy	4.30	0.702	16
Poor image associated with construction labour	4.30	0.693	16
Ever increasing sophistication	4.29	0.684	20
Lack of an adequate wage policy	4.29	0.647	20
Workers are prompted to pursue other career	4.27	0.698	22
Absence of regulations guiding minimum wage	4.20	0.669	23
Absence of workers full loyalty	4.16	0.858	24
Low influx of youths into the industry	4.13	0.785	25
Migration of construction workers	4.13	0.697	25
Inability to join trade union	4.08	0.932	27
Fewer workforce to replace the aging workforce	3.96	0.915	28

mean scores include the inability to join trade union (mean = 4.08), fewer workforce to replace the aging workforce (mean = 3.96), migration of construction workers (mean = 4.13), and low influx of youths into the industry (mean = 4.13). The standard deviation values show that the responses were relatively consistent across the sample population for most of the key factors. However, some factors such as lack of interest in technical skill acquisition, aging workforce, and absence of regulations guiding minimum wage had relatively higher standard deviation values. This suggests that the responses were more diverse for these factors.

Table 11.3 KMO Bartlett's test of sphericity for CSAS factors

Kaiser–Meyer–Olkin measure of sampling adequacy		0.860
Bartlett's test of sphericity	Approx. Chi-square	3226.472
	Df	378
	Sig	0.000

11.4.1 Key Factors Responsible for the Shortage of Skilled Artisans in the Nigerian Construction Industry (CSAS)

The results of the KMO Bartlett's test of sphericity for the CSAS factors are presented in Table 11.3. The Kaiser–Meyer–Olkin measure of sampling adequacy, with a value of 0.860, suggests that the sample size utilised in this analysis is satisfactory. The Bartlett's Test of Sphericity yielded a statistically significant result (χ^2 = 3226.472, df = 378), suggesting that the correlation matrix does not resemble an identity matrix. Consequently, it can be inferred that the variables under consideration are not independent of each other. The utilisation of factor analysis is justified in order to ascertain the latent components that contribute to the variability observed in the CSAS factors.

Table 11.4 displays the communalities for the CSAS factors before and after extraction. Communalities measure the variance of each variable that can be explained by all the factors extracted. The initial communalities for all variables are 1.000 because they represent the total variance of each variable. After extraction, the communalities range from 0.379 to 0.740, providing an indication of the extent to which each variable's volatility may be accounted for. The data presented in the table indicates that the first component accounts for a greater proportion of variance in the majority of the variables when compared to the second factor. For example, the variable "Fewer workforce to replace the aging workforce" has a communalities value of 0.740 in the first factor, indicating that 74% of the variance in this variable can be explained by the first factor. Similarly, the variable "Qualified and competent craftsmen" has a communalities value of 0.652 in the second factor, indicating that 65.2% of the variance in this variable can be explained by the second factor. Overall, the communalities suggest that the extracted factors are able to explain a significant amount of the variance in the CSAS attributes, and can be used to identify underlying factors that contribute to the challenges faced in the construction labour market.

The presented table displays the cumulative variation accounted for by the CSAS factors through the utilisation of principal component analysis. The eigenvalues prior to extraction correspond to the original eigenvalues, but the sums of squared loadings after extraction represent the amount of variation explained by each component. The rotation sums of squared loadings serve as a measure of the amount of variance that is accounted for by each component following the rotation process. The table shows that the first component explains 25.3% of the variance, the second component explains 8.43%, and the third component explains 6.45%. The first seven components together

Table 11.4 Total variance explained for CSAS factors

Component	Initial Eigenvalues			Extraction sums of squared loadings			Rotation sums of squared loadingsa
	Total	% of Variance	Cumulative (%)	Total	% of Variance	Cumulative (%)	Total
1	7.084	25.300	25.300	7.084	25.300	25.300	4.475
2	2.360	8.430	33.730	2.360	8.430	33.730	2.945
3	1.805	6.448	40.178	1.805	6.448	40.178	3.797
4	1.392	4.971	45.149	1.392	4.971	45.149	4.134
5	1.281	4.576	49.725	1.281	4.576	49.725	1.967
6	1.086	3.880	53.606	1.086	3.880	53.606	3.225
7	1.061	3.789	57.395	1.061	3.789	57.395	1.978
8	0.987	3.526	60.921				
9	0.923	3.298	64.219				
10	0.833	2.974	67.193				
11	0.755	2.698	69.891				
12	0.731	2.610	72.501				
13	0.700	2.500	75.002				
14	0.646	2.307	77.309				
15	0.624	2.228	79.537				
16	0.618	2.206	81.743				
17	0.586	2.091	83.834				
18	0.575	2.055	85.889				
19	0.546	1.950	87.839				
20	0.520	1.858	89.697				
21	0.444	1.584	91.281				
22	0.433	1.547	92.827				
23	0.410	1.463	94.291				
24	0.378	1.349	95.640				
25	0.341	1.218	96.858				
26	0.316	1.129	97.987				
27	0.306	1.094	99.082				
28	0.257	0.918	100.000				

Extraction method: principal component analysis
When components are correlated, sums of squared loadings cannot be added to obtain a total variance

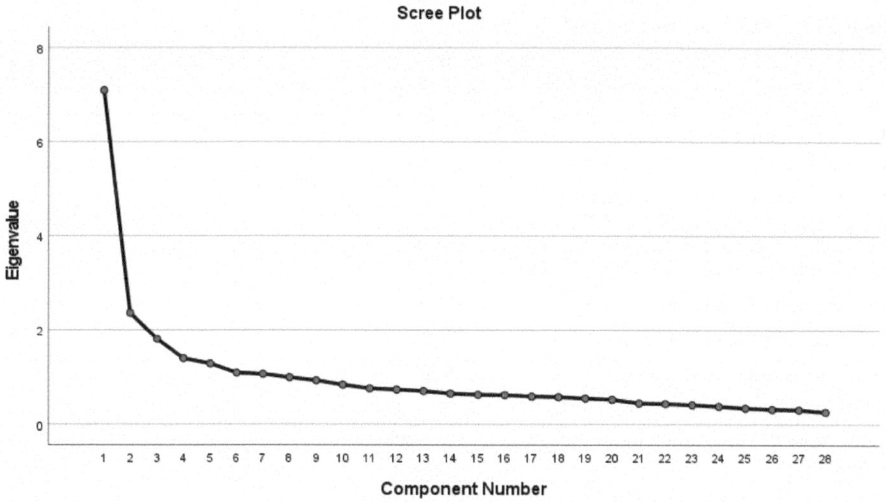

Fig. 11.1 Scree plot for CSAS Factors

explain 57.395 of the variance. The remaining components explain progressively smaller amounts of variance. The result was also represented in the Scree plot shown in Fig. 11.1.

Table 11.5 shows the pattern matrix for the factors identified in the CSAS (Construction Skills Assessment System) study. The factors are labelled as components 1 through 7. Component 1 includes factors such as lack of response to sporadic technologies, qualified and competent craftsmen, globalization in the industry, and ever-increasing sophistication. Component 2 includes factors such as inability to join trade union, absence of workers' full loyalty, low influx of youths into the industry, and fewer workforce to replace the aging workforce. Component 3 includes only one factor, which is the absence of regulations guiding minimum wage. Component 4 includes only one factor, which is migration of construction workers. Component 5 includes factors such as lack of an adequate welfare package and lack of an adequate wage policy. Component 6 includes factors such as focus on financial gains, inadequate training policy, inadequate discussion of workforce improvement, lack of staff and manpower development, absence of craftsmen training, high staff turnover, lack of well-structured incentives, and low demand for apprenticeship. Component 7 includes factors such as lack of interest in technical skill acquisition, workers are prompted to pursue other careers, and rework, cost, and time overrun. The extraction method used for this analysis is principal component analysis, and the rotation method used is oblimin with Kaiser normalization.

Table 11.6 shows the correlation component matrix for the CSAS factors. It shows the correlation coefficients between each of the seven components. Component 1 is positively correlated with components 3 and 5, but negatively correlated with component 4. Component 2 has a positive correlation with component 5, but a negative

Table 11.5 Pattern matrix for CSAS factors

	Component						
	1	2	3	4	5	6	7
Lack of response to sporadic technologies	0.835						
Qualified and competent craftsmen	0.793						
Globalization in the industry	0.746						
Ever increasing sophistication	0.691						
Inability to join trade union		0.815					
Absence of workers full loyalty		0.775					
Poor image associated with construction labour		0.405					
Low influx of youths into the industry			0.894				
Fewer workforce to replace the aging workforce			0.818				
Aging workforce			0.626				
Lack of interest in technical skill acquisition			0.562				
Less intelligent craftsmen							
Focus is on financial gains				– 0.755			
Inadequate training policy				0.668			
Inadequate discussion of workforce improvement				– 0.643			
Lack of staff and manpower development				– 0.591			
Absence of craftsmen training				– 0.531			
Absence of regulations guiding minimum wage					0.800		
Migration of construction workers					0.551		
Lack of an adequate welfare package						– 0.619	
Lack of an adequate wage policy						– 0.618	
High staff turnover						– 0.496	
Lack of well-structure incentives						– 0.456	
Low demand for apprenticeship						– 0.384	
Lack of motivation for improved productivity						– 0.364	
Lack of interest in skill acquisition							– 0.476

(continued)

Table 11.5 (continued)

	Component						
	1	2	3	4	5	6	7
Workers are prompted to pursue other career							– 0.469
Rework, cost and time overrun							– 0.385

Extraction method: principal component analysis
Rotation method: Oblimin with Kaiser normalization
Rotation converged in 19 iterations

Table 11.6 Correlation component matrix for CSAS factors

Component	1	2	3	4	5	6	7
1	1.000	0.140	0.300	– 0.398	0.078	– 0.254	– 0.221
2	0.140	1.000	0.115	– 0.126	0.166	– 0.211	– 0.127
3	0.300	0.115	1.000	– 0.301	0.150	– 0.231	– 0.096
4	– 0.398	– 0.126	– 0.301	1.000	– 0.084	0.259	0.164
5	0.078	0.166	0.150	– 0.084	1.000	– 0.083	– 0.030
6	– 0.254	– 0.211	– 0.231	0.259	– 0.083	1.000	0.126
7	– 0.221	– 0.127	– 0.096	0.164	– 0.030	0.126	1.000

Extraction method: principal component analysis
Rotation method: Oblimin with Kaiser normalization

correlation with component 4. Component 3 is positively correlated with components 1 and 5, but negatively correlated with component 4. Component 4 is negatively correlated with components 1, 2, and 3, but positively correlated with component 6. Component 5 has positive correlations with components 1, 2, and 3, but a negative correlation with component 6. Component 6 is positively correlated with component 4 and negatively correlated with component 5. Component 7 has a positive correlation with component 4 and negative correlations with components 6 and 7. These correlation coefficients can provide insight into how the different factors are related to each other.

11.5 Conclusion

Previous studies have identified various potential causes for the shortage of skilled artisans in the construction industry. This study, using South-western Nigeria as a case study, employs empirical methods and robust statistical analysis to estimate these causes and their relative contributions to the problem. The findings indicate that a lack of manpower development is the primary cause of the shortage of artisans in the construction industry. Consequently, this study proposes that stakeholders in the construction industry including construction firms, government entities, non-governmental organizations, and policymakers in the organized private sector should consider these validated factors when formulating and implementing capacity development programs aimed at enhancing the skills and abilities of artisans in the Nigerian construction sector.

Appendix

See Table 11.7.

Table 11.7 Communalities for CSAS factors

	Initial	Extraction
Fewer workforce to replace the aging workforce	1.000	0.740
Low influx of youths into the industry	1.000	0.731
Aging workforce	1.000	0.595
Lack of interest in technical skill acquisition	1.000	0.539
Absence of regulations guiding minimum wage	1.000	0.662
Migration of construction workers	1.000	0.560
Workers are prompted to pursue other career	1.000	0.532
Inability to join trade union	1.000	0.692
Absence of workers full loyalty	1.000	0.620
Lack of well-structure incentives	1.000	0.583
Lack of motivation for improved productivity	1.000	0.443
High staff turnover	1.000	0.661
Lack of an adequate wage policy	1.000	0.556
Lack of an adequate welfare package	1.000	0.520
Poor image associated with construction labour	1.000	0.525
Less intelligent craftsmen	1.000	0.379
Lack of interest in skill acquisition	1.000	0.540
Low demand for apprenticeship	1.000	0.548
Rework, cost and time overrun	1.000	0.468
Focus is on financial gains	1.000	0.583
Inadequate discussion of workforce improvement	1.000	0.528
Inadequate training policy	1.000	0.591
Absence of craftsmen training	1.000	0.537

(continued)

Table 11.7 (continued)

	Initial	Extraction
Lack of staff and manpower development	1.000	0.473
Ever increasing sophistication	1.000	0.543
Globalization in the industry	1.000	0.630
Lack of response to sporadic technologies	1.000	0.636
Qualified and competent craftsmen	1.000	0.652

Extraction method: principal component analysis

References

Aigbavboa CO (2014) An integrated beneficiary centered satisfaction model for publicly funded housing schemes in South Africa. Doctorate thesis, University of Johanessburg

Ajagbe AM, Ismail K, Aslan SA, Choi LS (2012) Investment in technology based small and medium sized firms in Malaysia: roles for commercial banks. Int J Res Manage Technol 2(2):147–153

Awe E (2006) Strategies for functional technical education and training of skilled craftsmen The builders' focus, 33–36. A magazine of the National Association of Building Students (NAOBS) Nigeria

Bilau AA, Ajagbe AA, Kigbu HH (2015) Review of shortage of skilled craftsmen in small and medium construction firms in Nigeria. J Environ Earth Sci 5(15):98–110

Carson B (2005) Capacity building strategies for sustainable development. In: National Construction Industry Council (ed) Proceedings of construction for sustainable development and environmental conservation, National Construction Industry Council, Malawi, vol 2 no 1, pp 119–129

Dantata S (2007) General overview of the Nigerian construction industry, Doctoral dissertation. Massachusetts Institute of Technology, Massachusetts

Dantong JSD, Lekjeb RS, Dessah E (2011) Investigating the most effective training for construction craftsmen that will optimize productivity in the nigeria construction industry

Dubem II, Stephen OO, Oluwaseyi AA (2012) An evaluation of personnel training policies of construction companies in Nigeria. J Hum Ecol 40(3):29–239

Ernst C, Sarabia M (2015) The role of construction as an employment provider: a world-wide input-output analysis. International Labour Organisation, Geneva. Experience. Available from: http://www.gla.ac.uk/centres/cradall/docs/Botswana-papers/

Fagbenle OI (2004) Factor affecting the performance of labour in Nigeria construction sites

Groot R, Molen P (2000) Workshop on capacity building in land administration for development countries: final report. ITC, The Netherlands

Kululanga GK (2005) Capacity building of construction contractors through collaboration. In: National Construction Industry Counci (ed), Proceedings of construction for sustainable development and environmental conservation, National Construction Industry Council, Malawi, vol 2 no 1, pp 130–138

Kululanga G (2012) Capacity building of construction industries in sub-Saharan developing countries. Eng Constr Archi Manage 19(1):86–100

Linstone HA, Turoff M (1975) The Delphi method: techniques and applications. Addison-Wesley Publishing Company, London

Medugu NI, Rafee Majid M, Bustani SA, Bala K, Abdullahi U, Mbamali I (2011) Craft skills availability in the Nigerian construction Industry: Perception of Contractors and Consultants. IUP J Infrastruct 9(3):63–73

Mwanaumo ME (2014) An integrated approach to multi-stakeholder interventions in construction health and safety, Doctoral dissertation. University of Johannesburg, Johannesburg

Offei-Nyako K, Osei-Tutu E, Fugar FDK, Adinyira E (2014) Skilled artisanal availability in the ghanaian construction industry. Covenant J Res Built Environ (CJRBE) 1(1):1–9

Ofori G (2001) Challenges of construction industries in developing countries: lessons from various countries. In: Proceedings of the 2nd international conference of the UB TG29 on construction in developing countries: challenges facing the construction industry in developing countries, Gabarone, pp 15–17 November 2000, 1–13

Ofori G (2012) Reflections on the great divide: strategic review of the book. In: Ofori G (ed) Contemporary issues in construction in developing countries. Spon Press, London

Okoye PU (2016) Optimising the capacity of Nigeria construction sector for socio-economic sustainability. Curr J Appl Sci Technol 1–16

Okuntade TF (2014) Building construction technician training: it's relevance to modern construction industry in Nigeria. Int J Technol Enhancements Emerg Eng Res 2(3):58–68

Olowa OW (2018) Determinants of rural residential solid waste collection services in Lagos State. Int J Sustain Dev World Policy 7(1):1–7

Oni OJ (2020) A strategic model for mitigating construction workforce shortages in South Western Nigeria. Int J Constr Eng Manage 9(3):99–106. https://doi.org/10.5923/j.ijcem.20200903.03

Onuka AOU, Ajayi, Kassim O (2012) Reflects of manpower development on workers job performance. Eur J Educ Stud 4(3):423–432

Sackman H (1974) Delphi assessment: expert opinion, forecasting and group process. Rand Corporation, Santa Monica, CA

Solomon O, Hashim HN, Mehdi BTZ, Ajagbe MA (2012) Employee motivation and organizational performance in multinational companies: a study of Cadbury Nigeria Plc. Int J Res Manag Technol 2(3):303–312

South African Construction (2015) Highlighting trends in the South African construction industry

Theisohn T, Lopes C (2011) Ownership leadership and transformation: can we do better for capacity development, 1st edn. Routledge. https://doi.org/10.4324/9781315065687

Ugheru DC (2006) Training of craftsmen for Nigeria construction industry. J Niger Assoc Eng Craftsmen 5:9–10

Yamoah EE (2014) The link between human resource capacity building and job performance. Int J Hum Resour Stud 4(3):139

Chapter 12
Unveiling the Causes of Business Failure in Ghana's Construction Industry: An Analytical Study

Alex K. Eyiah, Francis K. Bondinuba, and Evans Tee

Abstract This paper investigates the factors affecting the survival of construction businesses in Ghana, with recommendations for enhancing their continuous industry participation. The paper gauges contractors' and professionals' views on factors contributing to Ghana's construction business failure. It adopts a questionnaire survey distributed to contractors and industry professionals in Ghana. With an average response rate of eighty (80)%, the data was organised and analysed using mean scores of the identified variables. Paired samples T-test was also used to assess the significant difference between the two variables. Six common causes of business failure were identified: payment delays, excessive debt incurred, high charges on credit, working in unfamiliar localities, low profit margins and change in type/scope of work. It also emerged that the causes of business failure vary by the different financial classes of construction businesses operating in the industry. The study has practical, academic and policy implications for the survival and growth of construction businesses in Ghana. It is recommended that construction businesses should be strategic in selecting the type of work to undertake, the client group to work for, and the location to undertake projects. Government should expedite processes leading to establishing a Construction Industry Development Authority.

Keywords Business failure · Small and medium construction firms · Contractors · Construction industry · Ghana

A. K. Eyiah
Centre for Infrastructure Development in Africa, Manyo Plange Street, Accra, Ghana

F. K. Bondinuba (✉)
Department of Building Technology, Kumasi Technical University, Kumasi, Ghana
e-mail: francis.kbondinuba@kstu.edu.gh

E. Tee
Department of Business Administration, University of Professional Studies, Accra, Ghana

S. P. Akinbogun et al. (eds.), *Facility Management practices*,
https://doi.org/10.1007/978-3-031-65321-6_12

12.1 Introduction

Globally, the construction industry is often seen as a driver of economic growth. The industry can utilise local human and material resources to develop and maintain housing and infrastructure to promote employment and improve economic efficiency (International Labour Organization 2009; Erol and Unal 2015). The importance of the construction industry also stems from its strong linkages with other sectors of the economy. For example, energy, water and sanitation, banking, agriculture, ports and harbours, aviation, education, and manufacturing depend on some activities to flourish. In Ghana, the significant contribution of the construction industry is well documented (Owoo and Lambon-Quayefio 2018). Construction businesses, on their part, are the main channel through which the needed infrastructural facilities are provided. Despite such importance, construction business failures have been reported severally worldwide (cf. Jones 2019; Bal et al. 2013; Jagafa and Wood 2012; Mofokeng 2012; Strischek and McIntyre 2008). Failure is common in construction businesses, but their situation tends to be exacerbated due to the nature and characteristics of the construction industry. The immediate and ripple effects of construction business failure affect the whole economy (Abd-Hamid et al. 2015; Bal et al. 2013; Holt and Edwards 2012; Wong and Ng 2010; Tserng et al. 2011).

Sharma and Mahajan (1980) suggest that the extant literature should systematically study business failures and document evidence to enhance knowledge and reduce occurrences. In this regard, many models have been used to study the phenomenon (Onakoya and Olotu 2017). Among these, the Resource-based View (RBV) and the Organisation Ecology (EO) theories have been used extensively to explain the phenomenon of business failure (Arasti 2011). They are perfect lenses through which this study could be understood. Building on the arguments generated from these perspectives, other studies have emphasised the need to identify the factors leading to business failures to clarify the phenomena further and influence policy and industry responses (Assaad and El-adaway 2020). Given the significant economic contribution of the construction industry, some studies on construction business failure have been conducted to understand the phenomenon (cf. Holt 2013; Mahamid 2011; Osama 1997; Kivrak and Arsslan 2008). The lacuna found in these studies is that most tend to solicit only the views of construction firms seeking to identify the causes of failure. More so, whilst construction businesses can be placed under various categories depending on the type of work they undertake and, in some cases, their capacity, most studies lump them together as one group. There is also a need for more data on construction business failure in the Ghanaian context. This study is an attempt to bridge these gaps. It aims at identifying the common factors leading to construction business failures in Ghana. It also establishes whether there are differences in the category of a construction business and perceptions of factors leading to business failure. The remainder of the paper is structured as follows: concepts of business failure, resource-based view and organisation ecology theories, causes of business failure, research methods, results and discussions and finally, conclusions and recommendations.

12.2 The Concepts of Business Failure

The concept of business failure is viewed from different lenses throughout the literature. For instance, whereas Altman (1993) describes it in business insolvency, Richardson et al. (1994) deem it a shortage of money. Russell and Zhai (1996) describe it as termination in the business operations, whilst Appiah and Abor (2009) suggest it is a corporate failure. Ong et al. (2011) view it is about a business undergoing bankruptcy. Interestingly, a generic definition has been problematic such that (Cochran 1981) draws attention to the expanding and contracting nature of various definitions ranging from the dysfunctional broad to the acutely narrow. A popular view captured within the literature characterises failure as the cessation of involvement in a venture because it has not met and can no longer meet the minimum threshold for economic viability as stipulated by the entrepreneur (Ucbasaran et al. 2013). This definition explores failure from the entrepreneurship perspective, and as such, it is the entrepreneur's expectations of economic viability. In that context, Mbat and Eyo (2013) identified three types of business failure: low or negative returns, technically insolvent, and bankruptcy. According to them, low or negative returns would fail because there is no opportunity for expansion. Technically, insolvency refers to a business's inability to meet its liabilities as and when it is due. When the ratio of the total assets of the business and its liabilities is less than one for a given period, Mbat and Eyo (2013) explain that the business is said to be bankrupt.

Examination of various definitions revealed that most relate to the same significant event in the lives of businesses—the defining moment where it ceases to exist. It happens when the rate of return on invested capital is significantly and continually lower than what was invested. Cash flow is affected at a point the business needs to generate more profit, and financial obligations are challenging to meet. If this situation is not addressed, financial deterioration sets in and, eventually, business activities are disrupted. The final stage is when the business ceases to operate. To help place this paper into context, we adopt the above features to explain what we mean by construction business failure. By the term, we refer to entities that generally perform below the expected standard, compared with firms of similar capacities, for an extended period. The term also includes businesses that have ceased operating either for financially related reasons or willingly due to the owners failing to achieve their expectations.

12.2.1 Resource-Based View and Organisation Ecology Theories

The Resource-Based View (RBV) and Organisation Ecology (OE) theories have been used extensively in explaining business failure (cf. Godwin-Opara 2016; Amankwah-Amoah 2016; Barney and Hesterly 2012; Mellahi and Wilkinson 2004; Hannan and Freeman 1977). The arguments for the former centre on internal issues, and those for the latter relate to external environmental issues. The RBV explains that performance and sustainability depend on the availability of owned and controlled resources (Barney and Hesterly 2012). The categories of resources include physical, financial, technological, human and organisational (Ahmad 2015). The central argument of the supporters of RBV is that the availability of resources and capabilities provides the business with a competitive advantage, which helps it deliver superior customer service (Wu 2010). On the other hand, inadequate resources and capabilities make it difficult for the business to respond to operational challenges effectively. Thus, it is more likely to exit the marketplace as it cannot satisfy customer needs. In contrast, proponents of the organisation ecology theory argue that business failure relates more to the evolution of the external environment (Hannan and Freeman 1977). A company fails due to industry-specific and environmental factors, and managers cannot influence the firm's destiny (Mellahi and Wilkinson 2004). The external environment informs the shape of the venture, its operations, its ability to mobilise critical resources, circulation and the consumption of products or services (Venter et al. 2008). It exerts indirect pressure on the business (Cronje et al. 2004; Nieman and Nieuwenhuizen 2009), and because managers have little control over the turn of events, the likelihood of failure is increased (O'Connor 1994; Stoeberl et al. 1998). The external environment's elements vary from country to country and are primarily influenced by economic, political, and social conditions (Arasti 2011).

While the RBV and EO arguments present relatively straightforward answers, the reality of global competition makes these answers more complex and murkier. Also, one needs to indicate precisely which approach is more critical. In this sense, the true determinants of business survival and failure would comprise a combination of two forces (internal and external factors).

12.2.2 Causes of Construction Business Failure

The literature points to a myriad of root causes of business failures. Some studies argue that businesses fail due to strategic reasons such as the death of an owner without having a succession plan and one company's acquisition by another (Kuruppu et al. 2003; Mata et al. 2011). Others associate business failure with poor strategic leadership, lack of due diligence (Dasgupta and Sanyal 2010), lack of success (Teng et al. 2011), a series of unlikely events (Elenkov and Fileva 2006), economic crisis (Ong et al. 2011) inadequate capitalisation (Davidson and Maguire 2003). High levels of corruption, high credit costs, lack of access to finance, unfavourable regulatory regimes, high taxes rates increases in the interest rate and less demand for products and services have also been reported (Eyiah and Bondinuba 2020). Other factors associated with business failure include lack of industry experience, uncontrolled growth, marketing deficiencies, inadequate control systems, lack of entrepreneurial skills, lack of supervision, financial indiscipline, family interferences, low profit margins, bank loan delinquency, lack of commitment (Amankwah-Amoah 2016; Nangoli et al. 2013; Petrus 2009; Dun and Bradstreet 1986).

Within the construction industry, events that lead to business failure could be associated with the nature of the projects being executed and how well they have been handled. Davidson and Maguire (2003), based on their accountancy experience, identified the ten most common causes of construction business failure: obtaining work in a new geographic region; a dramatic increase in single job size; obtaining new types of work; growing too fast; high employee turnover; inadequate capitalisation; poor estimating; flawed accounting system; poor cash flow; and buying useless items. Schleifer (1990) identified ten similar activities as increasing project size; expanding into an unfamiliar location; undertaking new types of projects and between public/private sectors; replacing key personnel; not maturing in management as business expand; use of inadequate accounting systems; unprofitable projects; uncontrolled equipment costs; payment issues; and accounting systems. Strischek and McIntyre (2008) classified the causes into four groups: financial causes involving slow collections, low profit margins, insufficient capital and excessive debt, bank lines of credit and estimating and bid spread; management weaknesses involving inadequate management, poor risk management, changes in ownership or critical personnel, poor project management, and lack of a business plan; overexpansion involving a change in type/scope of work, the rapid growth of project size, new geographical location, inadequate accounting, estimation and equipment, and unfamiliar client; and uncontrollable involving unreasonable owners, inflation and high material costs or shortages, lack of skilled labour, and uncontrollable work environment. Other studies attribute the failure of construction businesses to how growth has been managed. Usually, this situation is preceded by events including growing beyond the financial and human resources of the firm; focusing on volume instead of profit; poor project selection; poor business acumen and poor leadership and succession (Craig 2012).

Empirical findings support the above observations (cf. Karas and Srbová 2019; Kivrak and Arslan 2008). Lower profit margin entities generally performance in Australia (Barbosa et al. 2017). As a result of fierce competition, construction businesses priced lower than expected. Using the wrong forms of finance negatively impacted their business (Barbosa et al. 2017). The phenomenon in the Czech Republic's construction industry was attributed mainly to high indebtedness, low labour productivity and poor working capital management (Karas and Srbová 2019). In the Turkish construction industry, lack of industrial experience and know-how and unfavourable economic conditions were the two significant issues causing business failure. Consistent with the findings in Australia, construction businesses bid extremely low to be successful in contract awards (Kivrak and Arslan 2008). In the West Bank in Palestine, Enshassi et al. (2006) found a delay in payment, high interest rate, lack of capital, cash flow problems, lack of industry experience and low profit margins. In their study of the USA, Arditi et al. (2000) found that over eighty (80)% of failures were caused mainly by insufficient profits, industry weakness, heavy operating expenses, insufficient capital, and excessive debt. In Saudi Arabia, difficulty in acquiring work, lack of experience, cash flow difficulties, and low profit margins were critical factors causing construction business failures (Osama 1997).

12.3 Research Design and Methods

The specific research that needs to be identified in this study provides information and a better understanding of the failure phenomenon within the construction industry in Ghana. The objective is to define an opinion held by key stakeholders to identify the common contributing factors to construction business failure in Ghana. On this basis, the descriptive approach was considered the most appropriate for the study. This approach afforded the researchers, to a large extent, the opportunity to obtain a high representation of the respondents' opinions. Observation, case study and survey are the data collection methods employed for this approach (Borg and Gall 1989). The survey was selected because secondary data sources of business failure are limited, especially in developing countries, as failed businesses disappear and owners of such ventures rarely talk about the reasons that led to failure (Shepherd 2003; Liao 2004).

The target population comprised construction businesses licensed by the Ministries of Roads and Highways (MRH) and Works and Housing (MWH). In effect, two groups of construction businesses were involved (road and building contractors). These groups are further categorised by the respective ministries, based on their resource capacities, class 1, class 2, class 3 and class 4 (Moo and Eyiah 2019). Consistent with Enshassi et al. (2006), the study sought the views of construction businesses on factors leading to business failure. Professionals within the construction industry are very much abreast with the activities and challenges of construction businesses. Due to needing a comprehensive sample frame, the study adopted a non-probability sampling technique. Specifically, the snowballing sampling technique was utilised to build the sample size through a line of references (Denscombe 2007; Tengeh et al. 2011).

The factors selected for the design of the questionnaires were those identified from the literature, which reflected the thinking of the resource-based view and organisation ecology theory and were considered empirically feasible. The author's experience in the construction industry also had a bearing on the selection of the factors. The questionnaire was designed using the Likert scale; one (1) to five (5). 1 = least significant and 5 = very significant. It was designed so respondents could list additional factors not captured from the literature. Industry experts and some contractors reviewed the questionnaire for refinement, and the revised version was piloted among a selected number of construction businesses and professionals, which minimised any structural weaknesses or deficiencies in the instrument. This practice improves content validity. The questionnaires had four sections: demographics; causes of failure, symptoms of failure; and survival factors. Finally, two questionnaires were developed: one for professionals and the other for construction businesses. A total of two hundred and fifty (250) questionnaires were distributed for answers; two hundred for construction businesses and fifty (50) for construction professionals.

Data from the survey were organised and analysed into summaries. The average or mean scores of the various variables were the critical parameter computed for the analysis. It was computed by adding all the constructs and dividing them by the number of measurements. This measure of the average scores used to rank the various variables is a sensitive measurement since its value always reflects the actual contributions of each of the variables that the respondent considered to the underlying factors influencing the business failures of the construction businesses. To determine if there was a mean difference between the two sets of observations, we used the paired sample t-test statistical procedure. This method is often used to compare two data sets. Specifically, it can be used to compare construction business failure variables. By analysing the data using this method, we were able to determine if there was a significant difference.

12.4 Analysis and Discussions

This section presents the average scores with the corresponding rankings of the factors causing construction business failure in Ghana. Tables 12.1 and 12.2 illustrate the analysed factors from the views of construction firms and professionals, respectively. From the contractor's perspective, the top five ranked factors causing construction business failure are excessive debt, delayed payments, high cost of credit, working in unfamiliar locations and low profit margin on projects. On the other hand, construction professionals ranked the causes of significance from delays in payments, excessive debt, change in the type of work, high cost of credit and demise of the company owner.

Following the computation of the average mean scores, the variables were compared among the two groups. The paired samples T-test was used to assess whether there was a significant difference between the two groups' views. From the results presented in Table 12.3, both groups agreed on almost all the factors except three as the causes of construction business failure. There were significant differences in the views of contractors and the professionals at a 5% level of significance that the death of the company owner, poor company organisation and lack of competent technical staff were causes of business failure.

The factors that the two groups disagreed on were ignored, and the top five position ranks from each group were considered the significant causes of construction business failures in Ghana. In general, six significant causes of business failure are identified as: delays in payment, excessive debt, high cost of credit, working in unfamiliar regions/localities, low profit margins and change in the type of work. Delayed payment and the high cost of credit can be classified as external factors. These exert pressure on the business (Cronje et al. 2004; Nieman and Nieuwenhuizen 2009), and because construction businesses have little control, the likelihood of failure is increased (O'Connor 1994; Stoeberl et al. 1998). On the other hand, the other factors (working in unfamiliar regions, low profit margin, excessive debt and change in the type of work) occur as a result of decisions made by management. These are internal matters; associated with the resource-based view theory. As a result of poor managerial and technical capabilities, the need to not adequately respond to business challenges.

Table 12.1 Descriptive statistics contractors

Business failure factor	N	Mean	Std. dev	Rank
Lack of experience in construction	120	3.22	0.881	14th
High turnover of key personnel	117	3.35	0.823	7th
Death of company owner	119	3.34	0.969	9th
Poor company organisation	119	3.22	0.761	14th
Lack of continuous access to work	118	3.35	0.820	7th
Pilfering (stealing) at site	118	3.19	0.795	16th
The low profit margin on projects	119	3.45	0.870	5th
Poor cash flow management	118	3.31	0.822	10th
Bidding lower than required	119	3.24	0.799	13th
Inadequate planning on projects	117	3.25	0.798	12th
Lack of access to capital	117	3.39	0.840	6th
Undertaking many projects at a time	120	3.15	0.876	19th
Increased size of projects	119	3.19	0.895	16th
Change in the type of work	117	3.08	0.911	21st
Inadequate management to a commensurate growth	119	3.26	0.887	11th
Lack of competent management staff	119	3.14	0.866	20th
Lack of competent technical staff	119	3.03	0.730	22nd
Negative influence by owner's family members	120	3.18	0.869	18th
Delayed payments from clients	118	3.62	0.942	2nd
High interest on bank loan	118	3.59	1.006	3rd
Working in unfamiliar regions/localities	118	3.54	1.010	4th
Excessive debt	118	4.06	1.056	1st

Authors' construct (2019)

The identified factors concur with the findings of similar studies elsewhere. Three of them: delayed payment, high cost of credit and low profit margin, featured in the key factors causing construction business failure in the West Bank of Palestine (Enshassi et al. 2006). A study in the USA also identified excessive debt and low profit margins (Arditi et al. 2000). Studies in Saudi Arabia identified A lower profit margin (Osama, 1997). Low profit margins caused more than 50% of business failure cases. High indebtedness features in a factor causing the bankruptcy of construction businesses in the Czech Republic (Karas and Srbová 2019) and in the USA (Arditi et al. 2000).

The issue of delayed payment has been a matter of concern in the construction industry in many countries and has been a major contributing factor to poor project performance (Bajjou and Chafi 2018). Construction firms have long bemoaned as unjustifiable the event, which has resulted in the loss of credibility of most of them. More worrying, construction businesses feel they would be victimised if they seek redress from the law courts. The immediate effect on the survival of construction

Table 12.2 Descriptive statistics professional

Business failure factors	N	Mean	Std. dev	Rank
Lack of experience in construction	50	3.42	0.883	11th
High turnover of key personnel	50	3.24	0.870	20th
Death of company owner	50	3.66	0.823	4th
Poor company organisation	50	3.62	0.753	6th
Lack of continuous access to work	19	3.29	0.816	17th
Pilfering (stealing) at site	50	3.02	0.795	22nd
The low profit margin on projects	50	3.26	0.751	19th
Poor cash flow management	49	3.33	0.747	16th
Bidding lower than required	50	3.24	0.822	20th
Inadequate planning on projects	49	3.29	0.816	17th
Lack of access to capital	49	3.37	0.834	15th
Undertaking many projects at a time	50	3.38	0.805	14th
Increased size of projects	50	3.50	0.953	9th
Change in the type of work	50	3.96	4.305	3rd
Inadequate management to a commensurate growth	50	3.40	0.857	13th
Lack of Competent management staff	49	3.55	0.738	7th
Lack of competent technical staff	48	3.48	0.743	10th
Negative influence by owner's family members	49	4.41	0.864	12th
Delayed payments from clients	50	4.50	5.618	1st
High interest on bank loan	50	3.66	0.823	4th
Working in unfamiliar regions/localities	50	3.52	0.909	8th
Excessive debt	50	4.14	1.010	2nd

Authors' construct (2019)

businesses is cash flow difficulties and the inability to continue the execution of the projects. The ripple effect can be dire: locked-up of invested capital, damaged reputation, inability to pay subcontractors, wages and salaries, and creditors, including banks. Recruiting and retaining a quality workforce becomes difficult, affecting future planning. Not being able to service their loans increases the cost of credit as interest charges compound.

It stands to reason that delayed payment is at the root of construction business failures, negatively impacting profitability and growth prospects. When payments are delayed, construction businesses are likely to default on loans, interest charges compound and the overall cost of credit increases. It plunges the businesses into excessive debt, which leads to failure (Karas and Srbová 2019). Some apply the wrong forms of finance in their operations. Cost of credit increases due to the lack of capital assets to serve as collateral on bank loans (Eyiah and Cook 2003). The inability to service outstanding loans due to delayed payments means many would have poor credit ratings, further impacting the cost of credit. Contractors with loans that have

Table 12.3 Paired samples test

	Paired Differences		*t*	df	sig
	Mean	Std. dev			
Lack of experience in construction	− 0.194	1.238	− 0.942	35	0.352
High turnover of key personnel	0.222	1.149	1.160	35	0.254
Death of company owner	− 0.472	1.383	− 2.048	35	**0.048**
Poor company organisation	− 0.556	1.027	− 3.247	35	**0.003**
Lack of continuous access to work	0.139	1.199	695	35	0.492
Pilfering (stealing) at site	0.194	1.167	1.000	35	0.324
The low profit margin on projects	0.194	1.142	1.022	35	0.314
Poor cash flow management	− 0.056	1.040	− 0.320	35	0.751
Bidding lower than required	− 0.028	1.108	− 0.150	35	0.881
Inadequate planning on projects	− 0.028	1.028	− 0.162	35	0.872
Lack of access to capital	0.111	1.116	0.598	35	0.554
Undertaking many projects at a time	− 0.361	1.099	− 1.971	35	0.057
Increased size of projects	− 0.194	1.451	− 0.804	35	0.427
Change in the type of work	− 1.056	4.974	− 1.273	35	0.211
Inadequate management to a commensurate growth	− 0.139	1.355	− 0.615	35	0.543
Lack of competent management staff	− 0.278	1.256	− 1.327	35	0.193
Lack of competent technical staff	− 0.528	1.028	− 3.081	35	**0.004**
Negative influence by owner's family members	− 0.111	1.090	− 0.612	35	0.545
Delayed payments from clients	0.222	1.149	1.160	35	0.254
High interest on bank loan	0.028	1.362	0.122	35	0.903
Working in unfamiliar regions/localities	0.083	1.251	0.400	35	0.692
Excessive debt	0.056	1.413	0.236	35	0.815

Authors' construct (2019)

fluctuating interest rates may find it more challenging to repay their loans. Higher loan payments may lead to reduced profitability which, affects business survival.

Expanding into new geographic areas or projects are ways to grow their businesses. The phenomenon could also be attributed to competition within the industry, which lead to limited construction opportunities (Kivrak and Arslan 2008). Entering new speciality areas is complex and risky, exposing the firms to several potential liabilities. In such events, labour supply, subcontractor quality, and suppliers can be dramatically different, resulting in unforeseen costs and problems. The problems are exacerbated if controls are not adequate, and inadequate, and the company needs more resources, workforce, and know-how to successfully execute growth with low; lower entry to construction creates a saturated marketplace with heavy competition, shrinking profit margins and better business practices. Many construction firms bid low to survive, resulting in a low profit margin (Kangari 1988).

12.4.1 Construction Business Failure: Financial Class and Type of Work

The positions of the various contractors relative to their financial classes and the type of work they undertake were also considered in the analysis. From Table 12.4, the total average score for the financial class 1 group was 3.00, with building class 1 and Road class 1 recording 3.00. Financial class 2 and class 3 had average scores of 3.54 and 3.18, respectively. Though all these are below the average industry score of 3.62, financial class 1 scored the least in terms of considering delayed payment as a primary cause of business failure. Furthermore, building contractors recorded higher average scores than road contractors on delayed payment as a factor. This group of contractors are indifferent to delayed payment, a business failure factor in the Ghanaian construction industry.

Financial class 1 recorded a total average score of 4.11, with building class 1 and road class 1 recording 4.21 and 4.00, respectively, concerning excessive debt. Both class 2 and class 3 recorded a high average score of 3.71. Similarly, concerning the high cost of credit as a cause of business failure, the class 1 contractors recorded the highest average score of 3.69, followed by the class 2 and class 3 groups with average scores of 3.47 and 3.39, respectively. Again, the Unfamiliar regions and location results showed that the class 1 contractors recorded the highest average score of 4.00, followed by class 3 and class 2 groups with average scores of 3.40 and 3.30, respectively. It is important to note that the class 1 group recorded higher average scores in all these three cases than the industry averages.

The fifth observed cause of business failure was the low profit margin. The scores relative to the different financial classes of the contractors showed that Class 2 recorded the highest average score of 3.32, with building Class 2 and road Class 2 recording 3.47 and 3.17, respectively. The next higher score was recorded by the Class 1 group (3.25), while Class 3 recorded a minor average. The industry players

Table 12.4 Views from construction firms: financial class and type of work

Class	Class 1		Class 2		Class 3		Total count
	Ave. scores	Count	Ave. scores	Count	Ave. scores	Count	
Delayed payment							
Building	3.00	19	3.73	64	3.36	22	105
Road	3.00	2	3.34	29	3.00	14	45
Total	**3.00**	**21**	**3.54**	**93**	**3.18**	**36**	**150**
Excessive debt							
Building	4.21	19	3.83	64	4.05	21	104
Road	4.00	2	3.59	29	3.36	14	45
Total	**4.11**	**21**	**3.71**	**93**	**3.71**	**35**	**149**
Cost of credit							
Building	3.37	19	3.63	64	3.48	21	104
Road	4.00	2	3.31	29	3.29	14	45
Total	**3.69**	**21**	**3.47**	**93**	**3.39**	**35**	**149**
Unfamiliar regions and location							
Building	3.50	20	3.56	63	3.64	22	105
Road	4.50	2	3.03	29	3.15	13	44
Total	**4.00**	**22**	**3.30**	**92**	**3.40**	**35**	**149**
Low profit margin							
Building	3.00	19	3.47	64	3.00	22	105
Road	3.50	2	3.17	30	3.14	14	46
Total	**3.25**	**21**	**3.32**	**94**	**3.07**	**36**	**151**
Change in type of work							
Building	3.00	18	2.97	64	3.14	22	104
Road	3.00	2	3.86	28	3.86	14	44
Total	**3.00**	**20**	**3.42**	**92**	**3.50**	**36**	**148**

Source Authors' constructs 2019

also considered the change in the type of work as a factor of business failure in the construction industry. However, since it was ranked relatively low among the contractors, we have decided it is not necessary to discuss it. Figure 12.1 illustrates a summary of the highest ranking from the respective groupings. It could be observed that delayed payment ranked highest with financial class 2 and building contractors. Excessive debt, credit and unfamiliar locality/regions all ranked the highest with financial class 1 contractor three factors ranked highest with building, road and building contractors, respectively. Lastly, low profit margins ranked highest with financial class 2 and road contractors.

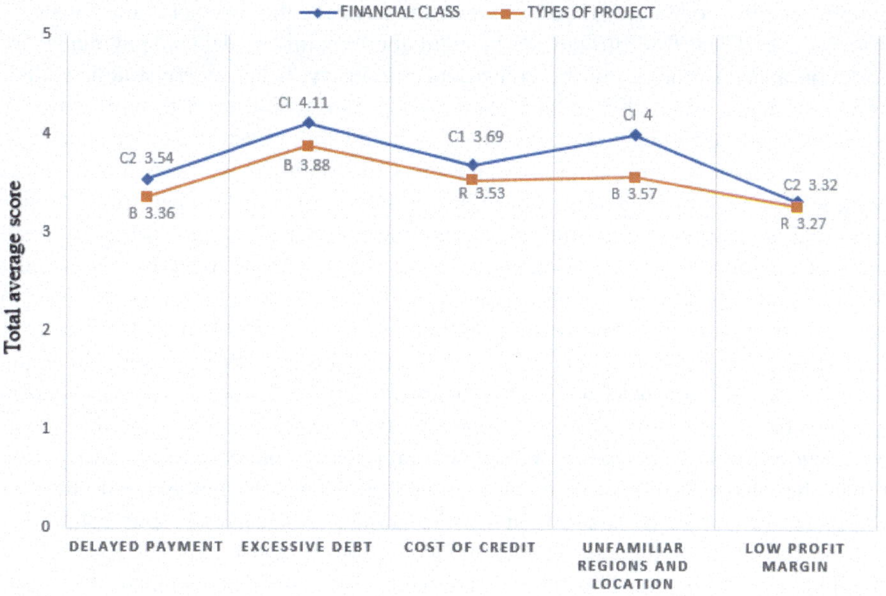

Fig. 12.1 Summary of highest-ranked factors (contractors' views)

12.5 Conclusion and Recommendations

It will most likely fail when a business cannot fulfil its financial obligations to stakeholders for an extended period. The most significant result from this study presents the significant factors leading to construction business failure in Ghana. Delayed payment, excessive debt, high cost of credit, unfamiliar project environment, low profit margins and change in scope of work are the leading causes of business failures. Four factors (excessive debt, low profit margin, delayed payment, and high cost of credit) are related to financial issues. The other two (working in unfamiliar regions and changing the type of work) are growth-related issues. Construction businesses need more control over two factors, delayed payment and high cost of credit. These are considered external environments associated with the organisation ecology theory. However, the other four factors can be regarded as internal factors, as they are based on the decision taken by management. It also emerged that the causes of business failure differ among the different financial classes of contractors and the type of project, with the financial class 1 category placing the most importance on the identified factors. This study's expanded knowledge and information can help guide entrepreneurs who want to enter and develop the construction industry in developing countries contexts. Two main implications emerged from the discussions. First, the study emphasises the importance of effective management and governance of construction businesses. These engender positive externalities that inure benefits culminating in good financial performance and ensuring the survival of the construction business. Cash-flow difficulties, brought about mainly by delayed payments, are common and unfortunate in the construction industry. It leaves construction businesses strategic in deciding which client group they would want to work with. A well-designed and managed marketing plan could link construction businesses to the private sector for more lucrative contracts. Construction businesses should select work within geographical areas that offer optimum cost control. They should maintain a business strategy that mitigates potentially negative impacts, primarily from increased competition, decreasing price levels, and high borrowing costs. They must pay close attention to overhead expenses as this is the tone of the surest way to increase profit margins. The second implication is closely linked with the role of policymakers in ensuring a conducive construction business environment. Given the importance of the construction business in the development process, serious attention needs to be given to this sector in supporting them to survive and grow their business. There would be a need for a local content policy to increase local construction firms' share in public construction procurement. It would also increase opportunities and incentives for local construction firms to develop their capacity, spur technology transfer and ensure sustainable management of public infrastructure. The government should expedite processes leading to establishing a Construction Industry Development Authority, which is strategic in strengthening, among others, the capacity of contractors.

Although the data analysed are likely to help provide valuable insight into the specific causes of construction business failures, the small sample size might be limited. A further study covering more subjects across the country would help expand the knowledge in this subject matter. A qualitative inquiry involving key stakeholders such as contractors' associations, chambers of construction, construction professionals and public construction agencies responsible for construction would also help provide an in-depth understanding of the subject of construction ilure in Ghana.

References

Abd-Hamid Z, Azizan NA, Sorooshian S (2015) Predictors for the success and survival of entrepreneurs in the construction industry. Int J Eng Bus Manag 7(12):1–11

Ahmad S (2015) Green human resource management: policies and practices. Cogent Bus Manag 2(1):1030817

Altman EI (1993) Corporate financial distress and bankruptcy: a complete guide top redicting and avoiding distress and profiting from bankruptcy. Wiley, New York, NY

Amankwah-Amoah J (2016) An integrative process model of organisational failure. J Bus Res 69(9):3388–3397

Appiah KO, Abor J (2009) Predicting corporate failure: some empirical evidence for the UK. Benchmarking Int J 16(3):432–44

Arasti Z (2011) An empirical study on the causes of business failure in Iranian context. Afr J Bus Manage 5(17):7488

Arditi D, Koksal A, Kale S (2000) Business failures in the construction industry. Constr Archit Manag 7(2):120–132

Assaad R, El-adaway IH (2020) Enhancing the knowledge of construction business failure: a social network analysis approach. J Constr Eng Manag 146(6):40–52
Bajjou MS, Chafi A (2018) Empirical study of schedule delay in Moroccan construction projects. Int J Constr Manag 1–18
Bal J, Cheung Y, Wu HC (2013) Entropy for business failure prediction: an improved prediction model for the construction industry. Adv Decis Sci 2013(14):1–14
Barbosa F, Woetzel J, Mischke J (2017) Reinventing construction: a route of higher productivity. McKinsey Global Institute
Barney J, Hesterly WS (2012) Strategic management and corporate competitive advantage: Concepts. Utah: Pearsons
Borg WR, Gall MD (1989) Educational research. White Plains, Longman. New York, NY
Cochran AB (1981) Small business mortality rates: a review of the literature. J Small Bus Manage 19(4):50
Craig P (2012) EU administrative law. Oxford University Press. Cape Town, SA
Cronje GJ, Du Toit GS, De KMA, Motlatla MDC (2004) Introduction to business management. Oxford University Press, Cape Town, SA
Dasgupta S, Sanyal D (2010) A stitch in time save nine: behind every business failurelies and untold story. Bus Strategy Series 11(2):1–6
Davidson RA, Maguire MG (2003) Ten most common causes of construction contractor failures. J Constr Account Tax 13(1):35–35
Denscombe M (2007) The good research guide. McGraw-Hill Education, Berkshire, England
Dun and Bradstreet (1986) Dun's census of American business. Dun and Bradstreet Corporation, New York, NY
Elenkov D, Fileva T (2006) Anatomy of a business failure: accepting the "bad luck" explanation vs proactively learning in international business. Cross-Cultural Manag 13(2):132
Enshassi A, Al-Hallaq K, Mohamed S (2006) Causes of contractor's business failure in developing countries: the case of Palestine 11(2):1–14
Erol I, Unal U (2015) Role of construction sector in economic growth: new evidence from Turkey. University Library of Munich, Germany
Eyiah AK, Cook P (2003) Financing small and medium-scale contractors in developing countries: a Ghana case study. Constr Manag Econ 21(4):357–367
Eyiah AK, Bondinuba FK (2020) Financing practices and preference of small and medium construction firms in Ghana. Int J Small Medium Enterprises Bus Sustain 5(1):36–60
Godwin-Opara M. N (2016). A Resource-Based Perspective on Financial Resource Strategies for Small Business Sustainability. (Doctoral Dissertation, Walden University)
Hannan MT, Freeman J (1977) The population ecology of organizations. Am J Sociol 82(5):929–964
Holt GD (2013) Construction business failure: conceptual synthesis of causal agents. Constr Innov 13(1)
Holt GD, Edwards DJ (2012) Innovation or business survival? A preliminary, qualitative study of UK construction plant supply chains. Constr Innov 12(1):99–122
International Labour Organization (ILO) (2009) ILO estimates and projections of the economically active population: 1980–2020
Jagafa K, Wood GD (2012) A turn-around model for construction company rescue: failure to success. In: Proceedings of engineering project organisation conference
Jones K (2019) 6 reasons why construction companies fail. Available [Online] at: https://www.constructconnect.com/blog/6-reasons-why-construction-companies-fail. Accessed 18 April 2020
Kangari R (1988) Business failure in construction industry. J Constr Eng Manag 114(2):172–190
Karas M, Srbová P (2019) Predicting bankruptcy in construction business: traditional model validation and formulation of a new model. J Int Stud 12(1):283–296
Kivrak S, Arslan G (2008) Factors causing construction company failure. Building Abroad, October, pp 297–305
Kuruppu N, Laswad F, Oyelere P (2003) The efficacy of liquidation and bankruptcy liquidation. Manag Int J 13(2):132–141

Liao J (2004) Entrepreneurship failures: key challenges and future directions. Entrepreneurship, pp 133–150
Mahamid I (2011) Causes of contractors' failure: contractor' view. In: 2nd international conference on construction and project management IPEDR, Singapore
Mata J, Antunes A, Portugal P (2011) Borrowing patterns, bankruptcy and voluntary models. J Constr Eng Manag 136(6):412–420
Mbat DO, Eyo EI (2013) Corporate failure: causes and remedies. Bus Manag Res 2(4):19–24
Mellahi K, Wilkinson A (2004) Organisational failure: a critique of recent research and a proposed integrative framework. Int J Manag Rev 5(1):21–41
Mofokeng TG (2012) Assessment of the causes of failure among small and medium- sized construction companies in the free state province, Doctoral dissertation, University of Johannesburg
Moo F, Eyiah A (2019) Factors influencing the growth of small and medium construction firms in Northern Ghana. J Afr Bus 1–16
Nangoli S, Turinawe DD, Kituyi GM, Kusemererwa C, Jaaza M (2013) Towards enhancing business survival and growth rates in LDC's. Int J Humanit Soc Sci 3(8):284–291
Nieman G, Nieuwenhuizen C (2009) The nature and development of entrepreneurship. Entrepreneurship. A South-African perspective, 2nd edn. Van Schaik, Pretoria, South Africa
O'Connor GC (1994) Differences in marketing strategies and operating efficiencies in surviving and failed organisations. J Strateg Mark 2(1):1–28
Onakoya AB, Olotu AE (2017) Bankruptcy and insolvency: an exploration of relevant theories. Int J Econ Financ Issues 7(3):706–712
Ong S, Yap VC, Khong RWL (2011) Corporate failure prediction: a study of public listed companies in Malaysia. Manag Finance 37(6):55–64
Osama MJ (1997) Reasons for construction business failures in Saudi Arabia. Proj Manag J 28(2):32–36
Owoo NS, Lambon-Quayefio MP (2018) The role of the construction sector in Ghana (No. 2018/119). WIDER Working Paper
Petrus HG (2009) An investigation into causes of success and failure in small businesses within the Department of Social Development in the Eastern Cape. Master thesis, Rhodes University, Grahamstown
Richardson B, Nwankwo S, Richardson S (1994) Understanding the causes of business failure crises: generic failure types. Manag Decis 32(4):922
Russell JS, Zhai H (1996) Predicting contractor failure using stochastic dynamics of option based credit models—models' performance and comparison with financial ratio prediction models for assessing going concern. Manag Audit J 18(6):577–590
Schleifer TC (1990) Construction contractors' survival guide, vol 66. Wiley
Sharma S, Mahajan V (1980) Early warning indicators of business failure. J Mark 44(4):80–89
Shepherd DA (2003) Learning from business failure: propositions of grief recovery for the self employed. Acad Manag Rev 28(2):318–328
Stoeberl PA, Parker GE, Joo SJ (1998) Relationship between organizational change and failure in the wine industry: an event history analysis. J Manage Stud 35(4):537–555
Strischek D, McIntyre M (2008) Credit risk management-red flags and warning signs of contractor failure-a timely review from a surety industry perspective on warning signs of impending contractor failure. RMA J 90(6):72
Teng HSS, Bhatia GS, Anwar S (2011) A success versus failure prediction failure model for small business in Singapore. Am J Bus 26(1):50–64
Tengeh HSS, Bhatia GS, Anwar S (2011) A success versus failure prediction failure model for small business in Singapore. Am J Bus 26(1):50–64
Tserng HP, Liao H, Tsai LK, Chen P (2011) Predicting contractor default with option based credit models—models' performance and comparison with financial ratio models. J Constr Eng Manag 136(6):412–420
Ucbasaran D, Shepherd DA, Lockett A, Lyon SJ (2013) Life after business failure: the process and consequences of business failure for entrepreneurs. J Manag 39(1):163–202

Venter R, Urban B, Rwigema H (2008) Advanced entrepreneurship. Oxford University Press, Cape Town SA

Wong JMW, Ng ST (2010) Company failure in the construction industry: a critical review and a future research agenda. Available [Online] at: www.fig.net/pub/fig2010/papers/ts01m%5Cts01m_wong_ng_4360.pdf. Accessed on 20 April 2020

Wu M (2010) The search for sustainable competitive advantage: a stakeholder management perspective, Doctoral dissertation, Massey University, Albany, New Zealand

Chapter 13
Towards an Understanding of Corporate Real Estate Management Practices in Ghana

Naana Amakie Boakye-Agyeman, John T. Bugri, Frank Gyamfi-Yeboah, Oti Amankwah, and Michael Adu-Kumi

Abstract Corporate real estate is a strategic resource generally undermanaged, particularly in the global south. This study examines the current approach to corporate real estate management (CREM) within five sectors in Ghana to explore whether the approaches differ across the sectors. A questionnaire survey supplemented by semi-structured interviews and document analysis was adopted. Descriptive statistics and logistic regression were used to determine if opinions on corporate real estate management practice differ by manager level, years of experience, and sector. Findings reveals that CREM practice is reactive in the five sectors with none of the organizations having a CREM policy that recognizes its importance in organisational activities and provides a structured approach to its management. The CREM function is thus not well integrated into corporate activities in the five sectors. The research has found that CREM is reactive and unsystematic CREM and corporate activities in the five sectors are not well aligned from corporate managers' point of view. It draws corporate policymakers' attention to the defective approach to CREM and the need for change.

Keywords Corporate real estate · Developing countries · Management practices · Ghana

---

N. A. Boakye-Agyeman (✉) · O. Amankwah
Department of Estate Management, Kumasi Technical University, Kumasi, Ghana
e-mail: naanaamakie@yahoo.com

J. T. Bugri · F. Gyamfi-Yeboah
Department of Land Economy, Kwame Nkrumah University of Science and Technology, Kumasi, Ghana

M. Adu-Kumi
Planning and Quality Assurance Directorate, Kumasi Technical University, Kumasi, Ghana

S. P. Akinbogun et al. (eds.), *Facility Management practices*,
https://doi.org/10.1007/978-3-031-65321-6_13

13.1 Introduction

For companies to achieve their goals, generate income, establish their identities in the community, and present themselves to the outside world, real estate is a crucial corporate resource (Khanna et al. 2012; Omar and Heywood 2014; Pfnuer et al. 2021). Again, accreditation or licensure in most corporate sectors, such as higher education, banking, security services, and health care, requires standard corporate real estate facilities (National Accreditation Board 2007; Bank of Ghana 2014; Bank of Ghana Act 2004; Oladokun and Aluko 2012; Van Reedt Dortland et al. 2013; Health Institutions and Facilities Act 2011; Ministry of Health 2014; Hopland and Nyhus 2015; Amos et al. 2019, Boakye-Agyeman et al. 2021). Well-managed corporate real estate (CRE) also contributes significantly to high-quality healthcare, higher education, banking, security services, and other services (Chotipanich and Lerteariyanum 2011; Povey and Peach 2013; Beckers et al. 2015; Van der Voordt 2016; Amankwah et al. 2019). These underscore the need for efficient corporate real estate management (CREM).

Most of the world's economies have reacted to the need for efficient CREM thus setting up CREM units in their organisations (Abdullah et al. 2011; Morri et al. 2017; Amos et al. 2019; Boakye-Agyeman et al. 2021). Empirical evidence shows positive change in top management's perception of the strategic value of CREM in developed economies (Allard and Barber 2003; Park and Glascock 2010; LaSalle 2015; Pfnuer et al. 2021), however, the same cannot be said of developing economies. CREM is relatively reactive in most developing countries (Tudzi et al. 2018; Amos et al. 2019; Boakye-Agyeman et al. 2021). In light of the COVID-19 pandemic in the context of the global south, addressing the reactive approach to CREM is imperative to changing the standards of practice (Amos et al. 2021).

Recent studies have shown that lack of CREM policy, CRE information management systems, inadequate authority, and unwelcoming corporate culture hinder proactive CREM practice, particularly in developing countries (Tudzi 2018; Amankwah et al. 2019; Amos et al. 2019; Boakye-Agyeman et al. 2021). CREM is usually excluded from top-tier corporate operations because senior management and other stakeholders undervalue its contribution to corporate success (Then 2005; Osei Kwarteng et al. 2012; Ijasan 2013; Boakye-Agyeman et al. 2021). CREM must justify its contribution to corporate success even in the developed world (Haynes and Appel-Meulenbroek 2014; Pfnuer et al. 2021).

According to Lalloo (2013), Tudzi (2018), and Amos et al. (2019), CREM practitioners have unsuccessfully tried to persuade senior management of the need for proactive CREM in cost-cutting measures. It has also been challenging to change the deeply established notion of the CREM position as taskmaster (Joroff et al. 1993), particularly in the global south and for that matter Ghana (Ijasan 2013; Ntene et al. 2020; Boakye-Agyeman et al. 2021). There are practical challenges One factor is the limited research on CREM approach in diverse sectors and its role in organisational processes in the global south especially, Ghana. Most research has concentrated on CRE branding, corporate and CREM strategy alignment and CREM value-adding characteristics (Khanna et al. 2012; Appel-Meulenbroek et al. 2010; Amos et al. 2015). Others (Ngwira et al. 2012; van Reedt Dortland et al. 2012; Beckers et al. 2015; van der Voordt 2016; Morri et al. 2017) focused on CREM practice in diverse sectors of developed economies with political and socio-economic characteristics dissimilar from those of developing economies like Ghana. In Ghana studies have concentrated on corporate real estate and facilities management in hospitals, universities, public buildings and banks (Tudzi 2018; Amankwah et al. 2019; Amos et al. 2019; Boakye-Agyeman et al. 2021). While these studies provide meaningful insights in understanding CREM practices, they are limited in establishing a holistic view of CREM practice in Ghana. Against this backdrop, this paper aims to examine CREM practices in five sectors in Ghana to explore whether the approaches adopted across the sectors differ. There are six parts to the paper. The study's origins and goals are discussed in Sect. 13.1. The understanding of CREM is explored in Part 2 from the viewpoint of the existing literature, and the study methodology is described in Part 3. The findings are reported in Part 4, and Sect. 13.5 brings the article to a close.

13.2 Overview of CREM Practice

According to Omar and Heywood (2014) and CoreNet Global (2015), CREM refers to physical fixed assets, such as land and buildings owned or leased for operational needs. According to the business's resource-based view (RBV), CRE is employed to establish and maintain a competitive advantage alongside capital and labour (Barney 1991; Heywood and Kenley 2008; Boakye-Agyeman et al. 2021). Research has shown that CRE accounts for between 20 and 50% of corporate assets worldwide and is the second-largest corporate expense (Jones and White 2012; Sulaiman et al. 2015; van der Beemt-Tjeerdsma and Veuger 2016; Boakye-Agyeman and Bugri 2019). As such, it must be carefully managed. CREM strives to integrate company's business and real estate strategy (Roulac 2001; Stadlhofer et al. 2012; Phelps 2010; Hayne 2012). Incorporating CRE strategy into the entire company plan offers value to the organisation by boosting customer satisfaction efficiency, and productivity (Scheffer et al. 2006; Boakye-Agyeman and Bugri 2019; Amos and Boakye-Agyeman 2022, 2023). In Table 13.1, several writers' definitions of CREM are presented.

The CREM definitions (Table 13.1) highlight the non-financial value adding attributes of CRE and the CREM function. According to Omar and Heywood (2010) and Hartmann et al. (2010), CREM performance is evaluated by operational efficiency rather than opportunities for value creation but the contribution of CREM to corporate success cannot be reduced to the success of a single performance mechanism (Lindholm et al. 2006; CoreNet Global 2015; Langford and Haynes 2015; Ivanov and Faulkner 2021). Although difficult to quantify, CREM impacts corporate image, credit rating, team member satisfaction, and productivity in addition to financial and measurable success (Scheffer et al. 2006; Appel-Meulenbroek et al. 2010; Ivanov and Faulkner 2021; Pfnuer et al. 2021). Firms with These extensive non-financial advantages need proactive CRE management.

Table 13.1 Summary of CREM definitions

Source	Definition/description
Dewulf et al. (2000)	The management of a corporation's real estate portfolio by aligning the portfolio and services to the needs of the core business (processes), to obtain maximum added value for the business and to contribute optimally to the overall performance of the corporation
Gibson (2000)	The economic, efficient, and effective acquisition, integration, coordination, and disposal of real estate resources to achieve ever-changing organizational goals
Kenley and Heywood (2000)	The management of property that is incidentally held, owned or leased by an organization to support its corporate mission
Krumm et al. (2000), Kaleva and Olkkonen (2001)	The management of an organization's real estate portfolio by aligning the portfolio and its related real estate services to the needs of the core business, to obtain maximum added value for the business and to contribute optimally to the overall performance of the organization
Lindholm (2008)	Concerns every real estate and facilities-related issue in public and private organization whose core business is not in real estate business
De Vries et al. (2008)	The range of activities undertaken to align CRE to the needs of the core business, to obtain maximum added value for the business and to contribute optimally to the overall performance of the corporation
Omar and Heywood (2010)	The real estate needs of non-real-estate organizations are met by CREM
Njungbwen and Udo (2010)	It concerns every real estate and facilities related issue in public and private organizations. In this case, the core business essentially is not real estate, but it can also exist in an organization that focuses on real estate business
Bakare (2010)	The strategic management of real estate held by corporations/ organizations, be it public or private, in pursuit of its primary business mission
Vermiglio (2011)	A systematic and multidimensional approach to the management of a real estate portfolio
White 2011	Property, people, place, and process managed together to optimize productivity
Ijasan (2013)	Deals with the management of a corporation's (enterprise's) real estate portfolio in both private and public sector organizations
Tudzi et al. (2015)	Deals with the proactive and holistic management of real estate and associated facilities held by corporate entities
CoreNet Global (2015)	Managing physical property for the business by acquiring, maintaining, and disposing of real estate throughout the life cycle of the property
Boakye-Agyeman and Bugri (2019)	Management of operational real estate assets for the achievement of corporate objectives

Source Authors construct

13.2.1 Proactive Corporate Real Estate Management (PCREM)

In the last four decades, there has been an increase in demand for a deliberate strategy for CREM. This is imperative considering the fact that proactive CREM would provide functional work environment to meet corporate challenges and ultimately enhance productivity, revenue growth and profitability (Lindholm and Leväinen 2006; Boakye-Agyeman et al. 2021; Amos and Boakye-Agyeman 2023). In spite of the importance of CREM to corporate success, its strategic importance is underrated in some organisations (Amos et al. 2015). Real estate is considered a cost center with focus on minimizing cost and reducing long-term financial exposure (Gibler and Lindholm, 2012). This is regrettable as real estate decisions have direct financial impact and also influence the core business of the organization (Roulac 2001; Amos et al. 2019). In particular, by aligning real estate decisions to corporate vision, real estate can contribute to organisational performance (Beckers et al. 2015; van der Voordt 2016). CREM must be well incorporated into business operations to be proactive. This includes strategic alignment through appropriate positioning of the CREM function within organisations, availability of a consistent CREM strategy, and a well-organised CREM unit linked with other functional units such as human resources, finance, information, and communication (Acoba and Foster 2003; Then 2005; Boakye-Agyeman and Bugri 2019). According to McDonagh and Nichols (2009) and CoreNet Global (2015), CREM must adopt a methodical strategy with sufficient backing from high management.

Proactive corporate real estate management (PCREM) requires clear strategic direction from senior management, adequate authority, updated CRE databases and supportive corporate culture (CoreNet Global 2015; LaSalle 2015; Boakye-Agyeman and Bugri 2019). This includes proper alignment of corporate and real estate strategies (Lindholm and Leväinen 2006; Langford and Haynes 2015) To achieve this, corporate real estate (CRE) managers should be able to formulate strategic plans that define the value-adding attributes of CRE that can enhance organizations' performance and stakeholders' wealth (Amos and Boakye-Agyeman 2023). There is also a need for a paradigm shift in the role of the CREM function in corporate activities from the traditional taskmaster to a business strategist role (Joroff et al. 1993; CoreNet Global 2015). Heywood and Kenley (2013) further recommended close relationships with senior management, CRE organisational structure and power, and CRE information management system as essentials of PCREM. Unfortunately, many firms, particularly those in the global south, see CRE as an expense that does not need proactive management (Amos et al. 2019; Boakye-Agyeman et al. 2021). Even where CREM is considered a strategic function, it must still validate its contribution to organisational success (Pfnuer et al. 2021). This is largely because corporate entities are deemed not to be in the real estate business (Langford and Haynes 2015; Ivanov and Faulkner 2021). Due to the competition for limited resources, particularly in non-real estate organisations, CREM, as a non-core function, must demonstrate its importance and added value to business performance in developing countries like

Ghana (Amos et al. 2019; Amos and Boakye-Agyeman 2023). Proactive management becomes essential in this context to ensure maximum contribution to warrant senior management's optimal support. Thus, the paper investigates the CREM approaches in five non-real estate "CRE-intensive" sectors in Ghana.

13.3 Research Methodology

The study employed an embedded mixed method where one data set provides a supportive, secondary role in a study based primarily on another data type (Creswell 2013). The largely quantitative study used a structured questionnaire supported by semi-structured interviews and document analysis to examine the CREM practices in five sectors in Ghana—tertiary education, health, banking, security services, and service industry, from the perspective of strategic (chief executives and directors), tactical (deputy directors) and operational (heads of departments) managers.

A questionnaire survey was used because it is generally economical for large data collection, reduces bias error, and allows for anonymity and generalisation (Denscombe 2010; Creswell 2013). The questionnaire and interview guide were developed based on an extensive literature review, the study's objectives, and the thematic areas such as CREM structure, the role of the CREM units, and CREM approach. The questionnaire was in three parts. Part one sought general information about the respondents and their sectors. Part two was on the role of CRE in corporate activities and three was on CREM practice. There were closed-ended questions with pre-set answers, such as yes/no, and multiple choice with an option for "other" that allows for responses not listed to be filled in by the respondents. The interview guide had open-ended probing questions that allowed respondents to give an in-depth explanation of the issues under consideration.

The multistage sampling procedure was used. The first stage involved the cluster sampling of five sectors where CREM is a requirement for accreditation. The purposive sampling was then used to select 35 organisations with CREM units using data from the Bank of Ghana, National Accreditation Board, Ghana Health Services, Ministries of Trade and Industry, Interior, and Defence. Organizations were selected from different sectors nationwide to ensure that findings represent views across the country. Again, purposive sampling was used to select business managers from the strategic, tactical, and operational levels. Studies have shown that corporate executives play important role in CREM decision-making (Teoh 1993; Tay and Liow 2006; Boakye-Agyeman et al. 2021). The respondents were purposely selected to solicit independent candid opinion of CREM practice especially from top-level managers of corporate entities who though not real estate professionals, necessarily must make key CREM decisions.

Questionnaires were sent to 1073 business managers; 680 were returned, giving a response rate of 63%. Out of these, 37 were incomplete and were therefore discarded, leaving 643 questionnaires for analysis. Additionally, semi-structured interviews were conducted with the 140 strategic-level managers. This allowed further probing to enhance, confirm and clarify findings from the survey (Saunders et al. 2012; Denscombe 2010; Creswell 2013). Official documents such as annual reports, strategic plans, organizational charts, statutes, and CREM policies obtained from 21 organizations were also reviewed to support or clarify the findings from the questionnaire survey and interviews with strategic managers. This allowed for triangulation and ensured the robustness of the results. Data was collected over one year. Anonymity was an ethical requirement of the responding institutions (non-disclosure of respondents and corporate identity). Descriptive statistics were used to summarize the views of respondents on CREM approaches in the five sectors. A logistic regression model was estimated to explore whether factors including years of experience, level of management and sector are related to respondents' views of CREM approaches.

13.4 Results and Discussion

Background of Respondents

Table 13.2 presents the level of management and years of experience of the respondents. The managers were drawn from three different levels of management, 22% strategic (top corporate decision making group-directors), 30% tactical (middle-level management- general managers), and 48% operational (first-level management-heads of departments) to give diverse view of the CREM practice. There is also a good representation of managers with a considerable level of experience as indicated by 48.3% of the respondents having more than 10 years of experience. This gives them good insight into CREM issues in the organisations.

13.4.1 Strategic Importance of CREM

This section gives an overview of the strategic role of CREM in organisational activities and the need for a proactive approach to CREM. Figure 13.1 shows that most business managers—83% each in tertiary education and banking, 76% for health,

Table 13.2 Background of respondents

Level of management	Years of experience					Total
	1–5	6–10	11–15	16–20	Above 20	
Strategic	16 (11%)	34 (24%)	19 (14%)	28 (20%)	43(31%)	140 (22%)
Tactical	42 (22%)	46 (24%)	44 (23%)	33 (17%)	29 (14%)	194 (30%)
Operational	118 (38%)	76 (25%)	53 (17%)	31 (10%)	31 (10%)	309 (48%)
Total	176 (27.4%)	156 (24.3%)	116 (18%)	92 (14.3%)	103 (16%)	643 (100%)

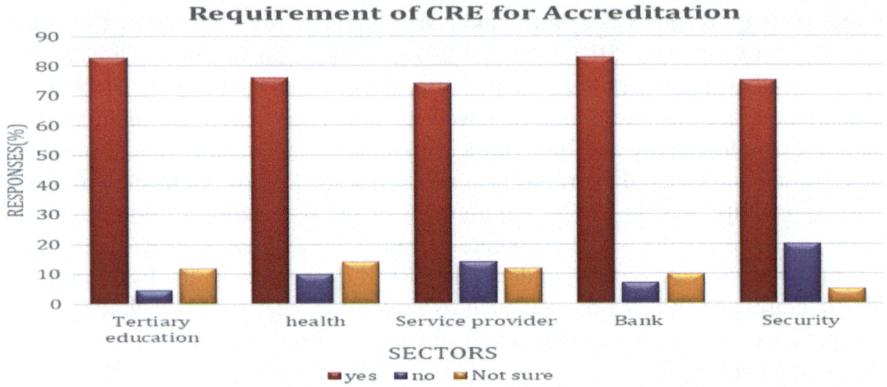

Fig. 13.1 Requirement of corporate real estate for accreditation. *Source* Field data

75% for security, and 74% for other service providers—indicated that CRE is a requirement for accreditation or licensing. This shows that organisations may not be able to start operations unless certain standard CREM facilities are in place. A Registrar explained that:

> Most of the programs submitted for accreditation have been put on hold because we do not have the required CREM physical infrastructure. The institution could not qualify to be upgraded to a Technical University because of inadequate physical infrastructure.

Document analysis also supported the findings as shown in the Health Institutions and Facilities Act, 2011(Act 829) Sect. 12(a) *(2 a-b)*

> There shall be attached to the application (a) the block plan of the facility for the practice; preliminary approval from the district planning authority of the District Assembly on the suitability of the land and facility to be used for the practice; 24 (1a-c) …… 'provide for the supervision and inspection of premises; provide for the prescription of safety standards for premises and sanitary arrangements of a practice; regulate the accommodation and appropriate equipment for each type of practice;

Aside from being a requirement for accreditation, the activities of the sectors are CREM-intensive and dependent. This confirms the findings of Allard and Barbers (2003), Appel-Meulenbroek et al. (2010), and Langford and Haynes (2015) that the contribution of CREM goes beyond the usual pecuniary benefit because the physical environment is a conduit for corporate teamwork. Thus, it is expected that CREM will be considered an important corporate function. The study, therefore, examined the corporate view of the CREM function as shown in Table 13.3.

13.4.2 Corporate View of the CREM Function

Table 13.3. shows that most of the managers—tertiary education, 226 (62%), health 68 (73%), banking 33 (55%), services 45 (52%) and security 30 (75%)—indicated

Table 13.3 Corporate perception of CREM function

Sector	Level of management		Corporate perception of the CRE management function				Total
			Taskmaster	Dealmaker	Controller	Business Strategist	
Tertiary education		Strategic	34	3	3	12	52
		Tactical	64	4	6	23	97
		Operational	128	9	22	55	214
	Total		226 (62%)	16 (4%)	31(9%)	90 (25%)	363 (100%)
Health		Strategic	22	2	0	5	29
		Tactical	26	2	2	4	34
		Operational	20	6	1	3	30
	Total		68 (73%)	10 (11%)	3 (3%)	12 (13%)	93 (100%)
Bank		Strategic	8	0	2	7	17
		Tactical	13	2	2	5	22
		Operational	12	0	2	7	21
	Total		33 (55%)	2 (3%)	6 (10%)	19 (32%)	60 (100%)
Service industry		Strategic	13	4	5	8	30
		Tactical	15	3	1	5	24
		Operational	17	5	0	11	33
	Total		45 (52%)	12 (14%)	6 (7%)	24 (27%)	87(100%)
Security services		Strategic	9	0	0	3	12
		Tactical	14	2	0	1	17
		Operational	7	1	2	1	11
	Total		30 (75%)	3 (8%)	2 (5%)	5 (12%)	40 (100%)
Total for all sectors		Strategic	86 (62%)	9 (6%)	10 (7%)	35 (25%)	140 (100%)
		Tactical	132 (68%)	13 (7%)	11 (5%)	38 (20%)	194 (100%)
		Operational	184 (60%)	21 (6%)	27 (9%)	77 (25%)	309 (100%)
	Total		402 (63%)	43 (7%)	48 (7%)	150 (23%)	643 (100%)

Source Authors construct

that the CRE management function is a taskmaster technical function (Joroff et al. 1993) and not strategic function that contributes to competitive advantage. It is worth noting that this view has not changed since Joroff et al. (1993) as 62% of strategic managers indicated that CREM is a taskmaster function. Strategic managers play a decisive role in CREM as observed in Van der Beemt-Tjeerdsma and Veuger (2016) and their view of CRE as a taskmaster function has serious implications. Various reasons were given for the foregoing findings. A Chief Executive Officer explained:

> The CRE unit and its staff need to be more proactive, you hardly get into contact with them, always waiting on people to report faults. During annual reviews, they hardly have anything to present, and no committee is constituted to review their activities.

In the banking sector, it was revealed that CRE is used in corporate branding, visibility, deposit mobilisation, branch expansions, and remodelling, confirming Appel-Meulenbroek et al. (2010) and Khanna et al. (2012). However, the role of the CREM unit is usually limited to basic maintenance and facilities management. A Director of Human Resources disclosed that:

> CRE is essential for corporate branding, so the CRE manager is a member of the branding team; however, CREM services end with furnishing the premises. The institution uses the Corporate Affairs unit for branding activities.

It can be deduced from the foregoing that the corporate perception of the CREM function is less favourable though CRE is considered a strategic corporate resource in the five sectors. This may have negative implications for the CREM approach as studies such as Schaefers (1999), McDonagh and Nichols (2009), and Boakye-Agyeman et al. (2021) have concluded that there is a strong correlation between senior management attitude and CREM practices.

Further investigations revealed a need for CREM policy and information management systems for proactive CREM. Table 13.4 shows that 94% of strategic, 93% of tactical, and 93% of operational managers in the five sectors said there were no CREM policies. Additionally, 76% of strategic, 79% of tactical, and 75% of operational managers indicated no CREM information management systems. Document analysis substantiated these findings:

> ... the expansion of health infrastructure continues to lag behind the real need. This may be due to the absence of Integrated Infrastructure Policy that would provide clear direction for managing health sector assets....
>
> As part of the Bank's new strategic vision, there is the need for a comprehensive policy on its fixed assets (corporate real estate). Even though various memoranda, circulars, and practices guide the way fixed assets are acquired, used, maintained, and disposed of, they have not been consolidated into a document to which reference can be made.

This contradicts Then (2005), and Boakye-Agyeman and Bugri's (2019) view that proactive CREM requires clear strategic direction and updated CRE databases for effective decision-making. This may be linked to the corporate perception of the CREM function (see Table 13.3). The development of CREM policy is evidently not a priority because the CREM function is technical with an indirect contribution to organisational performance as a Consumer Services Director indicated that:

Table 13.4 Availability of CREM policy and information management system

		CRE policy			CRE information management system			
Level of management	Sectors	Yes (No. (%)	No (No. (%)	Total (No. (%)	Yes (No. (%)	No (No. (%)	No response (No. (%)	Total (No. (%)
Strategic	Tertiary education	0 (0%)	52 (100%)	52 (100%)	2 (4%)	49 (94%)	1 (2%)	52 (100%)
	Health	1 (10%)	28 (90%)	29 (100%)	5 (17%)	23 (80%)	1 (3%)	29 (100%)
	Banking	4 (24%)	13 (76%)	17 (100%)	8 (47%)	9 (53%)	0 (0%)	17 (100%)
	Services industry	3 (10%)	27 (90%)	30 (100%)	11 (37%)	18 (60%)	1 (3%)	30 (100%)
	Security services	0 (0%)	12 (100%)	12 (100%)	4 (33.4%)	7 (58.3%)	1 (8.3%)	12 (100%)
	Total	8 (6%)	132 (94%)	140 (100%)	30 (21%)	106 (76%)	4 (3%)	140 (100%)
Tactical	Tertiary education	2 (2%)	95 (98%)	97 (100%)	6 (6%)	83 (86%)	8 (8%)	97 (100%)
	Health	1 (3%)	33 (97%)	34 (100%)	6 (18%)	27 (79%)	1 (3%)	34 (100%)
	Banking	5 (23%)	17 (77%)	22 (100%)	5 (23%)	14 (64%)	3 (14%)	22 (100%)
	Services industry	5 (21%)	19 (79%)	24 (100%)	4 (17%)	18 (75%)	2 (8%)	24 (100%)
	Security services	1 (6%)	16 (94%)	17 (100%)	2 (12%)	12 (71%)	3 (18%)	17 (100%)
	Total	14 (7%)	180 (93%)	194 (100%)	23 (12%)	154 (79%)	17 (9%)	194 (100%)
Operational	Tertiary education	15 (7%)	199 (93%)	214(100%)	20 (9%)	168 (79%)	26 (12%)	214 (100%)
	Health	0 (0%)	30 (100%)	30 (100%)	1 (3%)	24 (80%)	5 (17%)	30 (100%)
	Banking	2 (10%)	19 (90%)	21 (100%)	6 (29%)	11 (52%)	4 (19%)	21 (100%)
	Services industry	3 (9%)	30 (91%)	33 (100%)	11 (33%)	20 (61%)	2 (6%)	33 (100%)
	Security services	2 (18%)	9 (82%)	11 (100%)	3 (27%)	8 (73%)	0 (0%)	11 (100%)
	Total	22 (7%)	287 (93%)	309 (100%)	41 (13%)	23 (175%)	37 (12%)	309 (100%)
Grand total		44 (7%)	599 (93%)	643(100%)	94(15%)	491(76%)	58 (9%)	643 (100%)

Source Authors' construct

> After 50 years, the organisation has now acquired CRE management software that is yet to be installed. Though we appreciate the services of the CRE unit, especially for valuation services, CRE management itself is not a priority; it can be outsourced.

13.4.3 CREM Approach

The findings thus far show that (Fig. 13.1) aside from being a requirement for accreditation, organisational activities are highly CRE dependent. CRE must, therefore, be managed effectively and efficiently to support corporate objectives. However, Table 13.5 demonstrates that most managers (74%) in all the sectors indicated that CREM is reactive. This is confirmed by the lack of CREM policy, information management systems, and senior management perception. Contrary to Ntene et al. (2020), the lack of CREM policy indicates that corporate and CREM goals may be poorly aligned and the contribution of CREM to corporate performance overlooked.

Although many respondents across the sectors indicated that their organizations' CREM practice is reactive, there were differences in views on the extent to which the various sectors adopt PCREM approaches. A significant minority of respondents—44% in the health sector, 26% in the service industry, 23% in tertiary education and 22% each in security services and banking sectors indicated that CREM is proactive. To explore whether respondents' years of experience, level of management, and the sector of their organizations influence their views on the CREM approach, a logistic regression model was estimated with the results presented in Table 13.6.

A test of the full model shows that it is statistically significant at the 5% level X2(10, N = 643) = 39.95, $p < 0.001$. The significant coefficient for Health ($\beta = 0.981$, $p = 0.026$) confirms the earlier finding that respondents in the health sector were more likely to view the CREM as proactive compared to the other sectors. The odds of viewing the CREM as proactive are 2.668 times higher for the security services sector than other sectors. However, lack of CREM policy as stated earlier is prima-facie evidence that CREM is reactive in this sector as well. For the years of experience, which was coded using the "Above 20 years" as the reference group, only the 1–5 years of experience group ($\beta = 0.816, p = 0.008$) significantly impacted the view of the CREM approach. This suggests that less experienced managers with possibly limited knowledge of the operations of their organisations were more likely to view the management approach as proactive compared to more experienced business managers. Also, for the indicator variables for levels of management, only strategic ($\beta = 1.049$, p = 0.000) was significant, suggesting that business managers at the strategic level of management were more likely to view the CREM approach as proactive. The odds of such managers viewing the approach as proactive were 2.853 times more than those at the operational level.

Taken together, the results suggest that even though in practice, as has been demonstrated, the CREM approach adopted by the organizations is largely reactive, less experienced and managers at the strategic level may for different reasons be viewing the approach as proactive. It is quite telling that managers at the strategic level

Table 13.5 CREM approach

Sectors			CRE management approach		
			Reactive	Proactive	Total
Tertiary education		Strategic	35	17	52
		Tactical	78	19	97
		Operational	166	48	214
	Total		279 (77%)	84 (23%)	363 (100%)
Health		Strategic	14	15	29
		Tactical	17	17	34
		Operational	21	9	30
	Total		52 (56%)	41 (44%)	93 (100%)
Bank		Strategic	9	8	17
		Tactical	19	3	22
		Operational	19	2	21
	Total		47 (78%)	13 (22%)	60(100%)
Service industry		Strategic	18	12	30
		Tactical	18	6	24
		Operational	28	5	33
	Total		64 (74%)	23 (26%)	87 (100%)
Security services		Strategic	9	3	12
		Tactical	12	5	17
		Operational	10	1	11
	Total		31 (78%)	9 (22%)	40 (100%)
Total		Strategic	85 (61%)	55 (39%)	140 (100%)
		Tactical	144 (74%)	50 (26%)	194 (100%)
		Operational	244 (79%)	65 (21%)	309 (100%)
	Total		473 (74%)	170 (26%)	643 (100%)

Source Authors construct

who are expected to be better informed about the approach being used rather are more likely to erroneously consider it as proactive. Clearly, the lack of policy and information management systems are indication that there is no integration of both CREM and business information systems consistent with Then (2005) and Boakye-Agyeman et al. (2021). Consequently, there is likely to be no alignment of CREM and corporate strategies for top-down integration of business objectives, strategic planning for CREM, and bottom-up cross-functional integration of CREM and business units. Thus, strategic managers' view of CREM being proactive may be an indication of latent resistance to change of the prevailing reactive CREM approach in many organizations in Ghana as observed in Tudzi (2018) and Boakye-Agyeman and Bugri (2019). After all, change cannot take place if top management as the main policymaker does not see the need for it. CREM practice in Ghana is still custodial,

Table 13.6 Logistic regression relating views on CREM approach to experience, sector and management level

Independent variables	Coefficient	Standard error	*P*-value	Odds ratio
Sector				
Tertiary education	0.140	0.411	0.733	1.151
Health	0.981	0.441	0.026	2.668
Banking	– 0.043	0.500	0.932	0.958
Service providers	0.130	0.459	0.777	1.139
Years of experience				
1–5 years	0.816	0.309	0.008	2.261
6–10 years	0.210	0.315	0.505	1.234
11–15 years	0.112	0.338	0.740	1.119
16–20 years	0.237	0.346	0.494	1.267
Level of management				
Strategic	1.049	0.250	0.000	2.853
Tactical	0.336	0.227	0.140	1.399
Constant	– 1.983	0.461	0.000	0.138

Source Authors construct
Notes The variables Tertiary Education, Health, Banking and Service Providers are indicator variables coded as 1 for respondents in the given sector but 0 otherwise. The variables under "Years of Experience" are indicator variables coded as 1 if a respondent falls within a given experience category but 0 otherwise. The variables Strategic and Tactical are indicator variables coded as 1 for respondents in the given management level but 0 otherwise. Security services, above 20 years and Operational Management are the reference groups for Sector, Years of Experience and Level of Management respectively

unsystematic, and reactive. One major effect has been the unpreparedness of institutions in the management of unforeseen occurrences such as the Covid-19 pandemic. A structured approach to CREM is, therefore, worthwhile.

Langford and Haynes (2015) indicate that proactive CREM can reduce costs and contribute to corporate profit. But the study has revealed that CREM is reactive in the tertiary education, health, security, services industry, and banking sectors in Ghana. Although senior management acknowledges the important role of CREM in organizational activities, they see the CREM function as a technical task only for repairs and maintenance. However, given the CRE needs of the sectors, the huge capital investment in CREM and the insurgence of Covid-19, there is a need to follow the changing global trend.

13.5 Conclusion

Proactive CREM contributes to corporate success by providing conducive work environments that maximise productivity and stakeholders' wealth. This study examined the current approach to corporate real estate management (CREM) within five sectors in Ghana. Findings from the study show that CREM practice is reactive in the five sectors with none of the organizations having a CREM policy that recognizes its importance in organisational activities and provides a structured approach to its management. The corporate real estate management function is thus not well integrated into corporate activities. There is the need for a paradigm shift in the way CRE is managed and this would require a change in corporate perception of CREM. To achieve this, CRE managers must increase the understanding of CRE users on how CREM supports organisational activities through regular in-house education and well-planned CREM service provision.

Also, the core functions of CREM must be reframed. CREM must include more strategic real estate management activities such as CREM master planning that ensures effective space usage reduces utility and maintenance costs, improves organizations' financial performance, and reduces interruptions to core business. If CRE is managed proactively, with optimal senior management support, its contribution to corporate performance will move beyond physical 'brick-and-mortar space provision to intangible roles such as improving staff morale and productivity. From a policy perspective, the government would need to create a favourable legal and regulatory environment that encourages CREM policy formulation and implementation for improved efficiency in CREM.

Acknowledgements The authors want to convey their genuine appreciation to the management of all the institutions that granted permission for the study. We also appreciate the financial support of Kumasi Technical University for the Ph.D. work from which this paper is extracted.

References

Abdullah S, Razak AA, Pakir AHK (2011) The characteristics of real estate assetsmanagement practice in the Malaysian federal government. J Corp Real Estate 13(1):16–35

Acoba FJ, Foster SP (2003) Aligning corporate real estate with evolving corporate mission: process-based management models. J Corporate Real Estate 5(2):143–164

Allard LE, Barber C (2003) Challenges and opportunities in aligning real estate andthe workplace with business strategy: a survey of leading CEOs. J Corp Real Estate 5(3):213–220

Amankwah O, Weng-Wai C, Mohammed AH (2019) Modelling the mediating effect of health care healing environment on core health care delivery and patient satisfaction in Ghana. Environ Health Insights 13:1178630219852115. https://doi.org/10.1177/1178630219852115

Amos D, Gadzekpo A, Boakye Agyeman NA (2015) Real estate management and value-adding in corporate institutions in Ghana. Eur J Bus Manag 7(17):46–54

Amos D, Musa ZN, Au-Yong CP (2019) Performance measurement of facilitiesmanagement services in Ghana's public hospitals. Build Res Inf 48(2):218–238

Amos D, Musa ZN, Au-Yong CP (2021) Enhancing the role of facilities managementin the fight against the COVID-19 (SARS-CoV-2) pandemic in developing countries'public hospitals. J Facil Manag 19(1):22–31

Amos D, Boakye-Agyeman NA (2023) Content validation of value-adding indicators for corporate real estate management: insights from a developing country. J Corp Real Estate 25(1):77–97

Appel-Meulenbroek R, Havermans D, Jansssen I, Van Kempen A (2010) Corporate branding: an exploration of the influence of CRE. J Corp Real Estate 12(1):47–59

Bakare I (2010) Succinct introduction to corporate real estate management (CREM): Thebusiness perspective. Available at: https://ismaheelbakare.files.wordpress.com/2015/01/introduction-to-corporaterealestate-management-crem-bakare-2010.pdf. Accessed 31 Jan 2019

Banking Act 2004 (Act 673), Accra, Ghana

Bank of Ghana (2014) List of licensed banks, representative offices, and their registeredoffices in Ghana, available at: www.bog.gov.gh. Accessed 8 Aug 2015

Barney J (1991) Firm resources and sustained competitive advantage. J Manag 19(1):99–120

Beckers R, Van der Voordt T, Dewulf G (2015) Aligning corporate real estate with the corporate strategies of higher education institutions. Facilities 33(13/14):775–793

Boakye-Agyeman NA, Bugri JT (2019) Strategic corporate real estate management practice in Ghana. Prop Manag 37(3):432–449

Boakye-Agyeman NA, Bugri JT, Gyamfi-Yeboah F (2021) Challenges of strategic corporate real estate management practice in Ghana. Available online https://doi.org/10.1108/PM-10-2020-0065

Chotipanich S, Lerteariyanum V (2011) A study of facility management strategy: the case study of commercial banks in Thailand. J Facil Manag 9(4):282–299

CoreNet Global (2015) Essential guide to corporate real estate, CoreNet GlobalIncorporated, Atlanta

Creswell JW (2013) Research design: qualitative, quantitative, and mixed methodapproaches (4th edn). Sage Publications Inc., Thousand Oaks, California

Denscombe M (2010) The good research guide for small-scale social research project, 4th edn. Open University Press, Berkshire

De Vries JC, Hans de Jonge H, van der Voordt TJM (2008) Impact of real estateinterventions on organisational performance. J Corp Real Estate 10(3):208–223

Dewulf G, Krumm P, de Jonge H (2000) Successful corporate real estate strategies, Arko, Delft

Gibler KM, Lindholm A (2012) A test of corporate real estate strategies and operating decisions in support of core business strategies. J Prop Res 29(1):25–48

Gibson V (2000) Property portfolio dynamics: the flexible management of inflexibleassets. Facilities 18(3/4):150–154

Hartmann S, Linneman P, Pfnür A, Moy D, Siperstein B (2010) Responsibility for and performance of corporate real estate functions. J Corp Real Estate 12(1):7–25

Haynes BP (2012) Corporate real estate asset management: aligned vision. J Corp Real Estate 14(4):244–254

Haynes B, Appel-Meulenbroek R (2014) An overview of steps and tools for thecorporate real estate strategy alignment process. In: 21st annual European real estate society conference, Bucharest, Romania, 2014. Available at: http://shura.shu.ac.uk/9566/. Accessed: 8 Aug 2015

Health Institutions and Facilities Act, 2011, (Act 829), Accra, Ghana

Heywood C, Kenly R (2008) The sustainable competitive advantage model for corporate real estate. J Corp Real Estate 10(2):85–109

Heywood C, Kenley R (2013) Five axioms for corporate real estate management: a polemical review of the literature. Paper presented at the 19th Pacific rim real estate society conference, Melbourne, January 13–16. Available at: www.prres.net/. Accessed 8 Aug 2015

Hopland AO, Nyhus OH (2015) Does students satisfaction with school facilities affect exams results? Facilities 33(13/14):760–774

Ijasan KC (2013) Potentials of strategic corporate real estate management: South African perspective. Int J Econ Bus Finance 1(8):235–248

Ivanov SI, Faulkner M (2021) Corporate real estate ownership, cash and credit ratings. J Econ Stud 48(8):1460–1479

Jones K, White AD (2012) RICS Public sector asset management guidelines. RICS, Coventry

Joroff M, Louargand J, Lambert S, Becker F (1993) Strategic management of the fifth resource: corporate real estate. Industrial Development Research Foundation, Atlanta

Kaleva H, Olkkonen O (2001) Corporate real estate—international overview. KTIFinland: The Institute for Real Estate Economics

Kenley R, Heywood C (2000) Australian corporate real estate management identification of strategic issues in practice. Available at https://www.researchgate.net/publication/228900567. Assessed 19th Nov 2018

Khanna C, Van der Voordt TJM, Koppels PW (2012) Corporate real estate mirrors brand: a conceptual framework and practical applications. J Corp Real Estate 15(3/4):213–230

Krumm P, Dewulf G, De Jonge H (2000) What is corporate real estate? In: Dewulf G, Krumm P, De Jonge H (eds) Successful corporate real estate strategies. Nleuwegein, Arko

Lalloo A (2013) Corporate real estate practices in South Africa—a survey of the top 200 companies listed on The Johannesburg Stock exchange, M.Sc. Dissertation, University of the Witwatersrand

Langford L, Haynes B (2015) An investigation into how corporate real estate in thefinancial services industry can add value through alignment and methods ofperformance measurement. J Corp Real Estate 17(1):46–62

LaSalle JL (2015) Global corporate real estate trends: elevate to excellence. Available at: http://www.gcre.jonesslanglasallesites.com/. Accessed 14 Nov 2019

Lindholm A (2008) Identifying and measuring the success of corporate real estatemanagement. A doctoral thesis presented to the Helsinki University of Technology, Finland

Lindholm A, Leväinen KI (2006) A framework for identifying and measuring value added by corporate real estate. J Corporate Real Estate 8(1):38–46

Lindholm A, Gibler KM, Leväinen KI (2006) Modelling the value-adding attributes of real estate to the wealth maximization of the firm. J Real Estate Res 28 (4):446–475

McDonagh J, Nichols G (2009) Business strategy and property strategy—how strongis the linkage? J Corp Real Estate 11(4):213–228

Ministry of Health (2014) The health sector: medium-term development plan 2014–2017, Ministry of Health, Accra

Morri G, Djukic D, Chiavazza F (2017) Corporate real estate and performance: evidence from Italian manufacturing sectors. J Corp Real Estate 19(3):168–185

National Accreditation Board Act, 2007, (Act 744), Accra, Ghana

Ngwira MM, Parsa A, Manase D (2012) Effectiveness of property asset managementin Scottish councils. J Corp Real Estate 14(2):105–120

Njungbwen E, Udo GO (2010) Benefit of corporate real estate management tohigher education institutions. Department of Estate Management, University of Uyo, Nigeria. Available at http://www.academia.edu/710736/. Accessed 12th Sep 2014

Ntene T, Azasu S, Owusu-Ansah A (2020) Corporate real estate and corporate strategy alignment in South Africa. J Corp Real Estate 22(3):181–196

Oladokun TT, Aluko BT (2012) Corporate governance and real estate assetsmanagement in Nigerian banks. J Corp Real Estate 14(3):182–195

Omar AJ, Heywood C (2010) Corporate real estate management's credibilitypositioning status preliminary investigation of a branding model in practice. J Corp Real Estate 12(3):185–195

Omar AJ, Heywood C (2014) Defining a corporate real estate management's (CREM) brand. J Corp Real Estate 16(1):60–76

Osei Kwarteng K, Boadi-Siaw SY, Darko DA (2012) A history of the University of Cape coast: fifty years of excellence in tertiary education 1962–2012. University of Cape Coast, Cape Coast

Park A, Glascock JL (2010) The performance impact of strategic corporate real estatein franchise organisations. J Corp Real Estate 12(2):81–95

Pfnuer A, Seger J, Appel-Meulenbroek R (2021) Corporate real estate management and companies' success: empirical evidence for a conceptual framework. J Corp Real Estate 23(4):243–262

Phelps A (2010) Rationale, practice and outcomes in municipal property assetmanagement. J Corp Real Estate 12(3):157–174

Povey D, Peach N (2013) Understanding and implementing strategic asset managementat the University of Southern Queensland. Facilities 31(7/8):343–356

Roulac S (2001) Corporate property strategy is integral to corporate business strategy. J Real Estate Res 22(1–2):129–152

Saunders M, Lewis P, Thornhill A (2012) Research methods for business students, 6th edn. Pearson Education Limited, England

Schaefers W (1999) Corporate real estate management: evidence from German companies. J Real Estate Res 17(3):301–320

Scheffer JL, Singer BP, Van Meerwijk MCC (2006) Enhancing the contribution of corporate real estate to corporate strategy. J Corp Real Estate 8(4):188–197

Stadlhofer G (2010) Corporate real estate performance: contribution to core business competitiveness at global pharmaceutical enterprises. J Corp Real Estate 12(2):96–116

Sulaiman NN, Diah MLM, Ramin KA, Omar AJ, Ambar A (2015) Corporate real estate performance in Malaysian public listed companies. In: Paper presented at Pacific-rim real estate society conference, Kuala Lumpur, 18th–21st January 2015

Tay L, Liow KH (2006) Corporate real estate management in Singapore: a businessmanagement perspective 10(2):93–111

Teoh WK (1993) Corporate real estate asset management: the New Zealand evidence. J Prop Res 8(4):607–623

Then DSS (2005) A proactive property management model that integrates real estateprovision and facilities services. Int J Strateg Prop Manag 9(1):33–42

Tudzi EP, Gavu EK, Ayitey JZ, Boakye-Agyeman NA (2015) Corporate real estate management: a survey of literature. In: Badu E et al (Eds) Proceedings of 4th international conference on infrastructure development in Africa (ICIDA) conference Kumasi, Ghana, vol 25–26, pp 720–740

Tudzi EP (2018) Accessibility for persons with disabilities (PWDs) in corporate real estate management (CREM): a study of selected Universities in Ghana, PhD. Disseration, Kwame Nkrumah University of Science and Technology

Van Reedt Dortland M, Voordijk H, Dewulf G (2012) Towards a decision support tool for real estate management in the health sector using real options and scenario planning. J Corp Real Estate 14(3):140–156

Van den Beemt-Tjeerdsma A, Veuger J (2016) Towards a more professionalizedmunicipal real estate management. J Corp Real Estate 18(2):132–144

Van der Reedt Dortland M, Voordijk H, Dewulf G (2012) Towards a decision support tool for real estate management in the health sector using real options and scenario planning. J Corp Real Estate 14 (3):140–156

Van der Voordt TJM (2016) Adding value by healthcare real estate: parameters, priorities, and interventions. J Corp Real Estate 18(2):145–159

Vermiglio C (2011) Public property management in Italian municipalities: framework, current issues and viable solutions. Prop Manag 29(5):423–442

White AD (2011) A review of UK public sector real estate asset management. J Corp Real Estate 13 N

Chapter 14
Facility Management and the Quality of Teaching and Learning in Public Universities in Uganda

Eden Akategeka, Dennis Zami Atibuni, Fredrick Ssempala, and Denis Sekiwu

Abstract Facilities management arguably influences the quality of teaching and learning. In this chapter we undertake a conceptual review of the role of facilities management in the quality of teaching and learning in public universities across the world, but more specifically in Uganda. The chapter highlights the contextual and theoretical givens of quality of teaching and learning with respect to facilities management in the public university space. The review revealed a number of challenges of effective facilities management including limited funding, questionable staff practices, unaffordable maintenance costs, and delayed response times. Strategies for overcoming these challenges include upskilling administrative, support, and teaching staff to improve facilities management for fitness of purpose and hence promote the quality of teaching and learning. We argue that the quality of teaching and learning at public universities can be enhanced with better management of facilities.

Keywords Facilities management · Higher education · Massification · Quality · Teaching and learning

14.1 Introduction

The world generally and the world of academia in particular emphasizes the need for explicit attention to quality (IUCEA 2014). While quality is generally a difficult concept in itself, quality in higher education is much more confusing because in

E. Akategeka · D. Z. Atibuni (✉)
Busitema University, Tororo, Uganda
e-mail: zamidennis79@gmail.com; zamidennis.sci@busitema.ac.ug

F. Ssempala · D. Sekiwu
Kabale University, Kabale, Uganda

D. Z. Atibuni
Ali Mazrui Centre for Higher Education Studies, University of Johannesburg, Johannesburg, South Africa

S. P. Akinbogun et al. (eds.), *Facility Management practices*,
https://doi.org/10.1007/978-3-031-65321-6_14

Higher Education it is not always clear what the "product" is and who the "client" is. The notion of quality is of paramount importance because all education institutions need to train graduates who meet the needs of society. Besides, the labour market expects higher education institutions to provide the students with adequate knowledge, skills and favourable attitudes important for the right job fulfillment (IUCEA 2010). Nevertheless, the notion of quality is widely used in all educational contexts and increasingly so in the field of higher education. Most often the term is used in combinations such as quality systems, which is used to refer to systems encompassing different functions, aspects and levels of quality in an educational organization as a whole. Some studies of quality in education focus on the notion adoption of institutional evaluation (e.g., Hansen 2009), others on the concept of comparability (Kantelinen and Airola 2009), and yet others on the personal transformation of students or staff (Harvey and Newton 2007), while others have a marked emphasis on teaching and learning (e.g., Mikkonen et al. 2009).

Quality teaching and learning is comprised of two concepts, namely, quality teaching and quality learning. Quality teaching refers to the use of pedagogical techniques to produce learning outcomes for students (Cishe 2014) and quality learning, it refers to learning experience that causes a deep change in the students' outlook (Quinn 2015). In this study, basing on the propositions of Hénard and Roseveare (2012), quality teaching and learning is defined as referring to effective design of curriculum and course content, using a variety of learning contexts (including guided independent study, project-based learning, collaborative learning, and experimentation among others) soliciting and using feedback and effective assessment of learning outcomes. Elghany and Elharakany (2017) indicate that facilities in universities include provision centres, hostels, buildings, staff quarters, classrooms, laboratories, libraries, health centres and sport facilities among others. Managing the facilities mentioned on a high level of quality right from the planning level, through the development and implementation, may guarantee effective realisation of goals in educational institutions (Asiabaka 2008).

In Uganda, university education is experiencing a number of challenges which include among others; the increasing demand for university education, funding challenges, demography and quality of students admitted, competition, shortage of academic staff, ethical challenges, poor facilities and outdated curriculum, teaching and evaluation methods (Kasozi 2009). These have raised pertinent questions among managers of higher education institutions more than ever before with respect to how they are to meet national, regional and international integration requirements (Chaloff and Lemaitre 2009). Hence, interrogating the quality of teaching and learning in universities has become very paramount (Looney and Klenowski 2008), especially in the face of commodification and massification in higher education.

Massification in higher education is common throughout the developing and developed worlds and thus have been the subject of much debate among both policy makers and higher education researches. Higher education systems have moved from catering to elite and to mass state of massification and even post-massification in the western countries (Mok and Jiang 2016). The argument of massification of higher education all over the world has been that it reduces societal inequalities and makes education accessible to all social groups with in society (Kanyenyego 2009). This was because of the belief that increasing higher education enrolment would improve the quality of the population and enhance national competitiveness in the global world (Hawkins et al. 2014). The massification of higher education has resulted in the evaluation of social, cultural, economic, political and technological outcomes and advances as well as trends in the twenty-first century. The growth and evolution of these higher education systems presents challenges and opportunities. This chapter specifically maps the quality of teaching and learning in public universities in Uganda to the quality of facilities management, which facilities are grossly stretched and strained beyond capacity. First, we pitch a contextual understanding of the debate in the following section.

14.2 The Context of Facilities Management and Quality of Teaching and Learning

Recognising the importance of effective teaching and learning, the Government of Uganda in 2001 established the National Council of Higher Education (NCHE) as an agency to monitor the quality of teaching and learning in universities (Otto and Musinguzi 2013). The NCHE monitors the status of higher education institutions of learning, enrolments, academic programmes, academic staff profiles and physical infrastructure and facilities of the universities among others (Bailey 2014). The National Council for Higher Education (NCHE) requirements set out in Statutory Instrument 2005 Nos. 80 and 85 emphasize special attention to be paid to; access to relevant and up-to-date texts and other books as well as articles in journals; access to computer networks and the internet; access to general educational equipment, including power point projectors, slide projectors, overhead projectors, video, video cameras, access to sports and recreation facilities; and student exchange programmes to enhance student experiences (NCHE 2014).

Since the 1950s and 1960s to date, the nature of higher education has changed significantly: the elite university has changed into an institute of mass higher education. More and more students are being enrolled in higher education, and so creating pressure on national budgets. Expenditure per student has become much lower, but the government has to assure society that this does not endanger quality. Enrolment in higher education in Uganda was over 180,000 students according to Uganda Vice Chancellors' Forum (UVCF 2016), and about 87,120 students were in the nine public universities in Uganda (ESSAPR 2017). However, the capacity of universities to provide quality education in Uganda remains a very big challenge (Kasozi 2009). That is why Uganda is gradually losing its competitive edge in the region due to the inferior quality of its education (Kasozi 2010).

In the ranking of 200 best universities in Africa according to UniRank (2018), Makerere University ranked 1st and Mbarara University of Science and Technology

2nd in Uganda, but were ranked 38th and 185th respectively and were the only universities included in this ranking in Africa. Answers Africa (2018) ranked some public universities in Uganda worldwide as follow; Makerere University at the position of 2182, Mbarara University of Science and Technology at 5554, Kyambogo University at 7786, Gulu University at 9117 and Busitema University at 10,417. These rankings place public universities in Uganda at distant positions on the world scale that reflects a gap in the quality of education. The National Council for Higher Education (NCHE), the body established to oversee the implementation of the Universities and Other Tertiary Institutions Act 2001 in Uganda, has made efforts to put in place measures for quality assurance at institutional level that include the quality of teaching and learning and sufficiency of education facilities (NCHE 2014). Despite these efforts however, the management of facilities and the quality of teaching and learning still remains low in the public universities (NCHE 2014).

Facilities planning, facilities forecasting, facilities procurement, facilities installation, facilities maintenance, aspects of facilities management, may not be adequately done at the university due to its multi-campus model that strains the meager resources the university receives from the government (Kasozi 2017). This could affect the quality of the teaching and learning process that determines the quality of graduates in terms of the skills possessed (Allan 2006) and knowledge of the simple conceptual work taught (Kanyeihamba 2015). If the challenges mentioned above are not addressed, education which is an instrument of change will not ensure security, welfare and prosperity of the nation and its people (Sultana et al. 2016). However, there is a dearth of knowledge about facilities management and quality of teaching and learning at public universities in Uganda. This prompts the unanswered question as to whether the effectiveness of facilities management is a determinant of the quality of teaching and learning at public universities. We attempt to answer this question by providing a theoretical link between the two constructs.

14.3 Theoretical Grounding of the Facilities Management—Teaching and Learning Link

Henri Fayol's Administrative Theory

Henri Fayol (1841–1925) laid the foundations for modern scientific management also called principles of management as the underlying factors for successful management of organisations. Henri Fayol's administrative theory mainly focuses on the personal duties of management at a much more granular level (Olum 2004). Fayol's work is more directed at the management layer. Fayol believed that management had five principle roles: (a) to forecast and plan, (b) to organize, (c) to command, (d) to co-ordinate, and (e) to control. Forecasting and planning is the act of anticipating the future and acting accordingly. Organizing is the development of the institution's resources, both material and human. Commanding is keeping the institution's actions and processes running. Co-ordination is the alignment and harmonization of the group's efforts. Finally, control meant that the above activities were performed in accordance with appropriate rules and procedures. Fayol developed fourteen principles of administration to go along with management's five primary roles. These principles are: specialization/division of labor, authority with responsibility, discipline, unity of command, unity of direction, subordination of individual interest to the general interest, remuneration of staff, centralization, and scalar chain/line of authority, order, and equity, stability of tenure, initiative, and esprit de corps. Fayol's five principal roles of planning, organizing, commanding, coordinating, and controlling of management are still actively practiced today. Managers of organizations, in particular education institutions often use these principles to manage facilities in general so as to achieve organizational goals. Further still, the concept of giving appropriate authority with responsibility is well practiced, however, the other principles that include: discipline, subordination of individual interest, remuneration of staff, scalar chain or line of authority, equity and spirit de corps are consistently violated by most managers in organizations today (Olum 2004). This therefore explains why organizations and in particular Higher Education Institutions (HEIs) today still grapple with managerial challenges that affect the quality of education (Kasozi 2009). This theory was therefore not adopted for the study as it does not comprehensively address facilities management in relation to the quality of teaching and learning in Higher Education Institutions. Thus, the study adopted the Stage Theory for Higher Education Development.

The Stage Theory of Higher Education Development by Martin Trow (1970)

This study is underpinned by the Stage Theory of Higher Education Development propounded by Martin Trow (1972). The Stage Theory of Higher Education Development proposes that any huge or massive transition that is associated with growth will always have effect on the quality of teaching and learning because facilities are stretched (Liu and Mutinda 2016). Because facilities are stretched due to the ever increasing student enrolments amidst facilities that are not changing in proportion (Lin 2010), the Stage Theory of Higher Education Development defines quality as a change in the education concept, expansion of the function of education, diversity of teaching forms, change of the curriculum set up, teaching ways, entry requirements and management (Zhang and An 2010) and emphasizes the need for quality.

Quality constitutes the change of education concept, expansion of the function of education, diversity of teaching purposes and forms of Education, the change of curriculum set-up, the way of teaching, entrance requirement and management (Zhang and An 2010). However, quality teaching and learning requires not only adequate human resource but also sufficient and functional physical facilities and equipment (Goolam 2008). The Stage Theory of Higher Education Development shows that with massification in higher education, there is stress on facilities affecting the quality of teaching and learning. Therefore, this theory will be the basis for investigating how facilities management relates the quality teaching and learning. Accordingly, with massification there is a fundamental change from elite to mass, which does not only mean a sharp increase in the number of people who can receive higher education, but a change in quality, which plays an important role in the transition from elite to mass (Lin 2010). Elite higher education institutions are relatively small in size preparing the elite for polity and the learned professions. Their emphases are on the transmission of a general culture and lifestyle (Marginson 2010). However, massification, as a process challenges the traditional form of universities as centres of elite education where only a select few gain access affecting quality of teaching and learning (Hornsby and Osman 2014).

The Stage Theory of Higher Education Development explains quality as referring to the change of education concept, expansion of the function of education, diversity of teaching purposes and forms of Education, the change of curriculum set-up, the way of teaching, entrance requirement and management (Zhang and An 2010). However, quality teaching and learning requires not only adequate human resource but also sufficient and functional physical facilities and equipment. Insufficient physical facilities compounded by lack of maintenance, result in degradation of physical infrastructure because of massification. The progressive deterioration in lecture rooms and theatres, laboratories, library and administrative buildings affect the quality of learning (Goolam 2008). The growth of numbers without a parallel increase in state support threatens the quality. The Stage Theory of Higher Education Development shows that with massification in higher education, there is stress on facilities affecting the quality of teaching and learning. Therefore, we based our theoretical underpinning of the link between the quality of facilities management and the quality teaching and learning on this theory. In the section below, we attempt to isolate the historical and contextual perspectives of the quality of teaching and learning in higher education institutions.

14.3.1 Quality of Teaching and Learning in Education Institutions

Decades of pedagogical research have shown that what teachers do in classroom is undoubtedly the key educational determinant in student learning and achievement (Gauthier and Dembele 2004). Not all teaching practices can help pupils to achieve desired learning outcomes in the most effective way. There is a general rejection from stakeholders of what is referred to as "traditional" teaching because it's an essentially expository form of teaching, dominated by the teacher, which relegates pupils to a passive role, reduces their classroom activity to the memorization of data to be recited to the teacher, and in particular, leads to the acquisition of skills of a lower taxonomic level. The proposed alternatives to this form of teaching are the structured teaching approaches and discovery-based approaches that require adequate and up to date teaching and learning facilities.

Good teaching does not pay any benefit to teachers except self-satisfaction and also funding however, the requirements for quality teaching and are not available in universities (Sultana et al. 2009). Indeed, while a substantial number of studies on the effectiveness of teaching in developed countries conclude that the discovery-based teaching approach is weak, several studies conducted in developing countries appear to produce contradictory findings, presenting this approach as a factor of academic achievement (Gauthier and Dembele 2004). Every student enrolled in the institution of higher education dreams of a better quality of teaching and learning, and the university which provides quality teaching and learning is something they would be proud of. In higher education, university academic staff play an effective role to understand and to improve the teaching and learning process (Zaman 1998). Accordingly, improving the quality of teaching and learning in a University contributes to improved student learning and is likely to yield substantial dividends for students into the future (Chetty et al. 2011; Ladwig and Gore 2005).

The quality of teaching and learning enables the University to motivate the students better than the other institutions do (Suarman et al. 2013). Quality learning in relation to students encompasses what is learnt by the students, what is understood by them, what tasks they can perform, and the extent of competence the students acquire as the result of the interaction between the students and lecturers. The quality of teaching and learning relates to the practice of knowledge and learning experience obtained by the students (Chong and Crowther 2005).

Abba and Mugizi (2018) analysed the performance levels of academic staff in North West geo–political zones of Nigeria with staff from polytechnics as units of analysis. Descriptive findings revealed that teaching, supervision, research and publication and community service were effectively carried out but innovation was poor. The situation at Busitema University could be different due to inadequate funding from the government that affects the procurement of adequate teaching and learning facilities necessary to improve the quality of teaching and learning. Mwende (2014) investigated the school based factors influencing the quality of education in public secondary schools in Kitui County in Kenya. The findings revealed that inadequate physical facilities, staff and high fess affected the quality of teaching and learning. The findings in the study above reveal the same situation at Busitema University where libraries and staff are inadequate given the number of students admitted at the different campuses. This situation of inadequacy may have affected the quality of teaching and learning at Busitema University.

Suarman (2015) sought to identify the intermediary impact of the relationship between lecturers and students towards students' satisfaction and lecturers' teaching quality in a cross-sectional study in Indonesia. Descriptive findings revealed that a relationship between a lecturers and students is an important factor that determines lecturer's quality in teaching and learning quality in terms of the amount of knowledge inculcated, learning experience and satisfaction gained from students. This may be a challenge at Busitema University where the ratio of students to lecturers is still very high given the limited wage bill by the government for recruitment of more academic staff.

Szymenderski et al. (2015) investigated the impact of an assurance system on the quality of teaching and learning using staff of a university in Russia. The findings revealed that universities did not carry out quality teaching by effectively carrying out the role of informational, motivational and administrative functions to students. Lecturers were not interested in the quality of teaching and learning; they needed external control and were subsequently managed by making their remuneration dependent on the results achieved. The findings above reveal the same challenge the management of public universities face in motivating and coercing academic staff that may be affecting the quality teaching and learning. Barnett (2003) in his book 'Beyond all reason: Living with ideology in the university', argues that in this contemporary age of super-complexity the traditional assumption that the university is the embodiment of universal reason no longer holds. He argues that the old liberal idea that the university is a site of reason and therefore a 'public good' is being eroded by various ideologies and external forces that have permeated Higher Education. According to Barnett, such ideologies like: quality, manegerialism and entreprenueuralism have been imposed on or assimilated into Higher Education because society and governments no longer trust institutions of Higher Education.

Barnett (2003) stresses that Higher Education is now based on a range of different competing values as stakeholders of different interests try to impose their purposes in Higher Education. The state and the market have intruded into academic life in an unprecedented manner that has more often than not caused stress for both individuals and institutions. Governments have turned to Higher Education to provide the 'knowledge workers' that the knowledge based economies need. That's why governments are now interested in Higher Education for the socially useful knowledge. That it can deliver as a key component of national economic investment and as an instrument of economic mobilization and human resource development. This has therefore led to a drive for greater responsiveness and accountability on the part of Higher Education as governments seek to make the systems more efficient, effective and productive (Kathleen 2006). In developing countries like Uganda, as a result of economic constraints, learning conditions are not optimal and indicators relating to education provision, both quantitative and qualitative, are weak. In such contexts and in the light of the objectives of quality teaching and learning for all, both the question of educational effectiveness and that of efficiency must be raised. The next section of the study attempt to attribute the quality of teaching and learning to the effectiveness and efficiency of facilities management in the public universities.

14.3.2 Facilities Management in Education Institutions

Asiyai (2012) investigated school facilities management in public secondary schools in Delta State, Nigeria using teachers as units of analysis. Findings revealed that school facilities in the schools were generally in a state of disrepair. The findings further revealed that the roles of school administrators in the management and maintenance of school facilities included periodic inspection of facilities and decentralization of maintenance. The findings of Asiyai present a situation similar to that of Busitema University which inherited old and dilapidated teaching and learning facilities from the colleges it took over and whose periodic inspection of the facilities is rarely done. These issues could be having an effect on the quality of teaching and learning at Busitema University.

Lavy (2008) in a case study examined facility management practices in higher education buildings at a large southern university in the USA using students as the units of analysis. The findings revealed that there was effective facilities management because of organised facility management that improves the physical performance and appearance of a building and their systems and improved the efficiency with which the buildings were maintained and operated. However, the level of facilities management at Busitema University may not be as effective as that in the USA, a first world country for reasons related to inadequate funding, government policy, poor attitude of managers and inadequate managerial skills among others. These may have a great impact on the quality of teaching and learning at the University. Nik-Mat et al. (2011) in a study in Malaysia assessed the maintenance aspect of facilities management through a performance measurement system. The findings revealed that facilities management in Malaysia in terms of functional, technical and image were good. Accordingly, the effectiveness of facilities management service delivery to the end users was highly dependent on the facilities performance itself. Performance measurement (PM) is seen as a vital tool to assess the level of performance of facilities by analysing feedbacks from the end users. The findings above may differ with the situation at Busitema University where performance measurement of facilities (PM) may inadequately be done and hence affecting the quality of teaching and learning.

Pitt et al. (2016) examined customer satisfaction towards facilities and service quality in two private hospitals in Bangkok with the patients as the units of analysis. The findings revealed that overall facilities management was good with clients satisfied with facilities quality. This may not be the case with Busitema University, an education institution in Uganda, a third world country where the quality of facilities management may not be satisfactory to customers due to the poor management skills exhibited by most managers in particular those in Education Institutions which could be affecting the quality of teaching and learning. Xaba (2012) in a qualitative study analysed school facilities maintenance in South Africa with principals and deputies as the study sample. The results revealed that facilities maintenance in the schools was poor because schools generally did not have organisational structures for planned facilities maintenance, nor did they have policies on facilities maintenance. Facilities maintenance at schools mainly related to concerns with facilities repairs, (mostly as the need arises) and general campus cleanliness mostly with emergency and corrective forms of maintenance as opposed to crucial preventive maintenance. The results above reveal a situation that may be prevailing at Busitema University where facilities maintenance may be poor and a policy on facilities maintenance may be absent hence affecting the quality of teaching and learning.

14.3.3 The Facilities Management—Quality of Teaching and Learning Link Explicated

Different scholars (Akpabio 2015; McGowen 2007; Ndirangu and Udoto 2011) have analysed the relationship between facilities management and quality of teaching and learning. Akpabio (2015) carried out a review on managing school facilities for quality education in Nigeria looking at challenges and prospects. The review revealed that school facilities give meaning to the teaching and learning process. Therefore, facilities are an integral part of the overall management of public universities and university managers should use the diagnostic approach in carrying out comprehensive assessment of the facilities to determine areas of need. This requires integrated efforts of all stakeholders of the university who possess the expertise needed for accurate and up-to-date assessment of all aspects of teaching and learning facilities. Afework and Asfaw (2014) investigated the availability of school facilities and their impacts on quality of education in Eastern Hararge zone schools in Harari regional state in Ethiopia using school principals, district and regional education bureau heads. The results showed that school facilities and instructional materials were lacking therefore affecting the quality of teaching and learning. These findings reveal a situation existing at Busitema University where facilities seem to be inadequate which may mean that a negative effect on the quality of teaching and learning at the University.

Leung and Fung (2005) investigated the effectiveness of the enhanced school facilities and the impacts of this enhancement on the learning behaviours of students in schools in Hong Kong. The results revealed that enhanced school facilities improved learning behaviour with most of the improved facilities management correlating to enhanced learning behaviour. These results mean that if Busitema University improved its teaching facilities, it would probably enhance learning behaviour and hence improve the quality of teaching and learning. McGowen (2007) explored the possible relationship between school facility conditions and school outcomes such as student academic achievement, attendance, discipline, completion rate and teacher turnover rate using principal and principal's designee on high school campuses in Texas in the USA. Descriptive results revealed that facilities management affected the quality of teaching and learning influencing student achievement, attendance and completion rate. The results of the study reveal that if effective management of facilities at Busitema University, would positively influence the quality of teaching and learning and hence improve students' achievement, attendance and completion rate.

Ndirangu and Udoto (2011) reported findings on the perceptions of quality of educational facilities in Kenyan public universities and the implications for teaching/learning, and the learning environment using undergraduate students and academic staff from five public universities. The results revealed that the quality of the lecture facilities provided by Kenyan public universities did not meet quality measures of adequacy. Hence facilities management did not support quality teaching and learning. This may mean that if Busitema University does not improve the perception students and staff have on the quality of lecture facilities, it may affect the teaching and learning process and hence the quality of education.

14.3.4 Challenges Faced in Facilities Management

Scholars (e.g. Kamarazaly 2014, Kamarazaly et al. 2013; Wuni et al. 2017) investigated challenges faced in facilities management. Bin Syed Mustapa and Jusoff (2009) sought to determine the extent of facilities in Malaysia and reviewed the changes in the property sector that positively contributed to opportunities in facilities and the challenges that needed to be taken into account. The findings revealed that challenges of facilities management included; first, the non-existence of standards that can be used to measure the quality level and performance of both traditional and integrated FM applied by the building/property management. Second, the lack of local expertise to provide immediate response to failure of service as well as lack of facilities management practitioners in the local market that can provide advice or assistance in the implementation of facilities management. Third, implementation of facilities management was considered late for some properties and there were many aging buildings with high deterioration level. Fourth, low level software development specific to facilities management and adoption of integrated facilities management required high initial cost, unless the computerised programmes can be found locally in the market. The challenges mentioned above seem to be the same challenges at Busitema University affecting the management of facilities.

Kamarazaly (2014) analysed challenges in strategic facilities management looking at problems faced by university facilities managers in New Zealand and Australia with facilities managers in the Australasian universities as units of analysis. The results showed that the critical challenges facing the university facilities managers were to finance-related current internal challenge, economic-related current external challenges and sustainability-related future challenges. Kamarazaly et al. (2013) sought to identify the challenges faced by university facilities managers using members of the Australasian university facilities managers as units of analysis. The results showed that the critical challenges facing facilities management comprised inadequate funding, emergency management and business continuity planning, statutory compliance, sustainability and environmental stewardship, keeping up with rapid changes in technology, operational efficiency, identifying and meeting stakeholder needs, maintenance and manpower. The results though of university facilities in a developed country, still seem to portray the same challenges facing Busitema University.

Ogbeifun (2011) analysed facilities management in a multi-campus setting using staff of the University of the Witwatersrand in South Africa. The findings showed constraints impairing the effective facilities management were low capacity and ineffective use of the existing structure and available tools. Therefore, the University needed to complement their investment in capital developments with concerted efforts on capacity building; in terms of increase in staff numbers as well as retraining of existing staff. The findings of the study reveal the need for Busitema University to put in more effort in capacity building of its staff members if the quality of facilities management is to improve. Wuni et al. (2017) examined the causes and effects of poor facility management in public second cycle schools in the Upper East Region of Ghana. The findings revealed that poor facility management challenges included lack of professional managers, attitude of deferred maintenance, budgetary limit on maintenance expenditure, sidelined in the financial planning and capital budgeting of institutions, and non-privy of facility management staff to the building design and construction stage. The lack of professional managers and budgetary limits on maintenance expenditure seem to be the challenges Busitema University faces that should be addressed to improve its level of facilities management.

14.3.5 *Enhancing Facilities Management, Quality Teaching and Learning in Public Universities*

Based on the foregoing synthesis on facilities management and quality of teaching and learning, this study offer the following insights into enhancing the quality of facilities management so as to enhance the quality of teaching and learning at the public universities. First, managers of universities need to improve facilities management. This should be through increasing funding of facilities, ensuring that the number of students increases proportionately with the facilities, providing adequate supervision and increasing the number of staff. Second, the public university managers should work to solve challenges of facilities management. This should involve increasing the number of staff, training the staff in facilities management, ensuring fast handling of facilities that need to be taken care of, integrating technology in facilities management and increasing funding for facilities. The managers need to make effort to promote the quality of teaching and learning. This should be through promoting explorative techniques, offering courses that enable acquisition of job market skills, promoting teaching and learning that involves use various facilities and providing a teaching and learning environment is conducive.

14.4 Conclusion

The debate on facilities management and quality of teaching and learning at public universities reveals the former as a significant determinant of the latter. This agrees to the Stage Theory which claims that the availability of quality facilities and instructional materials positively impacts the quality of teaching and learning. The debate further reveals that facilities management is faced by several challenges which include: lack of sufficient management staff, low competence of staff in facilities management, delayed handling of facilities that need to be taken care of, lack of integration of technology in facilities management and limited funding. We therefore conclude that effective management of facilities relates to enhanced teaching and learning quality, and therefore the managers of institutions, in particular public universities in Uganda, need to consider effective management of facilities as a critical management function in order to achieve the institutional goals and objectives and hence fitness for purpose.

References

Abba HD, Mugizi W (2018) Performance of academic staff in polytechnics: an analysis of performance levels in North West geo–political zone of Nigeria. Arts Humanit Open Access J 2(3):198–203

Afework TH, Asfaw MB (2014) The availability of school facilities and their effects on the quality of education in government primary schools of Harari regional state and East Hararghe Zone, Ethiopia. Middle Eastern Afr J Educ Res 11:59–79

Akpabio IU (2015) Managing school facilities for quality education in Nigeria: challenges and prospects. J Qual Educ 11(1):1–6

Allan J (2006) The quality of graduates in the labour market: a study of graduate skills and employability. J Educ Work 19(2):141–157 https://doi.org/10.1080/13639080600675858

Asiabaka IP (2008) Effective realization of goals in educational institutions: a strategic planning approach. J Educ Adm 46(6):733–746 https://doi.org/10.1108/09578230810908242

Asiyai RI (2012) Assessing school facilities in public secondary schools in Delta State, Nigeria. Afr Res Rev 6(2):192–205

Bailey T (2014) The role and functions of higher education councils and commissions in Africa: a case study of the Uganda National Council for higher education. Centre for Higher Education Transformation (CHET), Wynberg, South Africa

Barnett R (2003) Beyond all reason: living with ideology in the university. Open University Press

Bin Syed Mustapa SAH, Jusoff K (2009) Facility management challenges and opportunities in the Malaysian property sector. J Sustain Develop 1(2):79–85

Chaloff J, Lemaitre G (2009) Managing highly-skilled labour migration: a comparative analysis of migration policies and challenges in OECD countries. OECD Directorate for Employment, Labour and Social Affairs, Paris, France

Chetty R, Friedman JN, Hilger N, Saez E, Schanzenbach DW, Yagan D (2011) How does your kindergarten classroom affect your earnings? Evidence from Project STAR. Q J Econ 126(4):1593–1660. https://doi.org/10.1093/qje/qjr041

Chong BK, Crowther F (2005) A new framework for measuring the quality of outcomes-based engineering education. In: Proceedings of 35th annual conference on frontiers in education. FIE'05, pp F1E–22. IEEE

Cishe EN (2014) Fostering quality teaching and learning in higher education through academic staff development: challenges for a multi-campus university. Mediterr J Soc Sci 5(25):272–277

Economic Sector Strategic Planning and Research (ESSAPR) (2017) Higher education in Uganda: status report

Elghany MMA, Elharakany RA (2017) Questioning facilities management in higher education in Egyptian context for better decision-making process of funding allocation. Int J Bus Manag 12(7):129–137

Gauthier C, Dembele M (2004) Quality of teaching and learning in developing countries: a review of the literature. Int J Educ Res 41(4–5):339–354 https://doi.org/10.1016/j.ijer.2005.04.007

Goolam M (2008) The effects of massification on higher education in Africa. Int J Afr Higher Educ 1(1):1–79

Hawkins JN, Mok KH, Neubauer D (2014) Higher education governance in East Asia: politics, policy, and markets Palgrave Macmillan

Harvey L, Newton J (2007) Transforming quality in higher education: a framework for assessment. In: Westerheijden DF, Stensaker B, Rosa MJ (Eds) Quality assurance in higher education: trends in regulation, translation and transformation. Springer, pp. 227–242

Hansen M (2009) Institutional evaluation: a framework for assessing quality in higher education. Qual Assur Educ 17(2):133–146. https://doi.org/10.1108/09684880910951343

Hénard F, Roseveare D (2012) Fostering quality teaching in higher education: policies and Practices. An IMHE Guide for Higher Education Institutions, pp 7–11

Hornsby DJ, Osman R (2014) Massification in higher education: large classes and student learning. High Educ 67(6):711–719

Inter-University Council for East Africa (2010) Skills and attitudes for the world of work: a report on the survey of employers and graduates in East Africa. IUCEA

Inter-University Council for East Africa (2014) Quality assurance in East African universities: status, challenges, and prospects. IUCEA

Kamarazaly MA (2014) Challenges in strategic facilities management: Analysis of problems faced by university facilities managers in New Zealand and Australia. A PhD of Massey University, Massey, Australia

Kamarazaly MA, Mbachu J, Phipps R (2013) Challenges faced by facilities managers in the Australasian universities. J Facil Manag 11(2):136–151

Kantelinen H, Airola M (2009) Comparability of higher education qualifications: a conceptual framework. High Educ 58(5):697–709. https://doi.org/10.1007/s10734-009-9225-6

Kanyeihamba GW (2015) Unemployable graduates employed in Uganda. Daily monitor. Available at: www.monitor.co.ug/Home/Magazines/People&Power. Retrieved 7 July 2018

Kanyenyego P (2009) Massification of higher education in Africa: challenges and opportunities. J High Educ Afr 7(2):1–15

Kasozi AB (2009) Challenges facing higher education in Uganda: poor facilities, outdated curriculum, teaching and evaluation methods. J High Educ Afr 7(1):1–18

Kasozi AB (2010) The quality of higher education in Uganda: challenges and prospects. J High Educ Afr 8(1):1–18

Kasozi ABK (2017) The national council for higher education and the growth of the university sub-sector in Uganda, 2002–2012. CODESRIA, Dakar, Senegal

Kathleen M (2006) Enhancing the efficiency, effectiveness, and productivity of educational systems. J Educ Adm 44(3):278–291. https://doi.org/10.1108/09578230610664814

Ladwig JG, Gore JM (2005) Improving the quality of teaching and learning in universities: a framework for action. High Educ Res Dev 24(2):47–165. https://doi.org/10.1080/07294360500062258

Lavy S (2008) Facility management practices in higher education buildings: a case study. J Facil Manag 6(4):303–315

Leung MY, Fung I (2005) Enhancement of classroom facilities of primary schools and its impact on learning behaviors of students. Facilities 23(13/14):585–594

Lin S (2010) The construction of the guarantee system of quality in graduate education. China Electric Power Edition 15:34–35

Liu Z, Mutinda G (2016) On progress of mass tertiary education: case of Lebanon, Kenya and Oman. Higher Educ Stud 6(4):1–11

Looney A, Klenowski V (2008) Curriculum and assessment for the knowledge society: interrogating experiences in the Republic of Ireland and Queensland, Australia. Curriculum J 19(3):177–192

Marginson S (2010) Higher education in the global knowledge economy. Procedia Soc Behav Sci 2(5):6962–6980

McGowen RS (2007) The impact of school facilities on student achievement, attendance, behavior, completion rate and teacher turnover rate in selected Texas high schools, PhD Thesis, Texas A&M University, Texas, USA

Mikkonen J, Ruohoniemi M, Lindblom-Ylänne S (2009) Teaching and learning in higher education: perspectives on faculty development. High Educ Res Dev 28(3):287–301. https://doi.org/10.1080/07294360902839933
Mok KH, Jiang J (2016) Massification and post-massification of higher education in Western countries: reflections on its implications for East Asia. J High Educ Policy Manag 38(3):248–262. https://doi.org/10.1080/1360080X.2016.1174409
Mwende GL (2014) School based factors influencing quality of education in public secondary schools in Kitui Country, Kenya, Masters dissertation, University of Nairobi, Nairobi, Kenya
National Council for Higher Education (2014) Requirements for higher education in Uganda. Statutory Instruments 2005 Nos. 80 & 85
Ndirangu M, Udoto MO (2011) Quality of learning facilities and learning environment: challenges for teaching and learning in Kenya's public universities. Qual Assur Educ 19(3):208–223
Nik-Mat NEM, Kamaruzzaman SN, Pitt M (2011) Assessing the maintenance aspect of facilities management through a performance measurement system: a Malaysian case study. Proced Eng 20:329–338
Ogbeifun E (2011) Facilities management in a multi-campus setting: a case study of the University of the Witwatersrand, Masters dissertation, University of the Witwatersrand, Johannesburg, South Africa
Olum Y (2004) Management duties and responsibilities: a detailed analysis. J Manag Res 4(2):123–145
Otto F, Musinguzi B (2013) The accreditation process and challenges of private religious based universities in Uganda. Int J Pedagogical Innov 209(1179):1–7
Pitt M, Chotipanich S, Issarasak S, Mulholland K, Panupattanapong P (2016) An examination of facility management, customer satisfaction and service relationship in the Bangkok healthcare system. Indoor Built Environ 25(3):442–458
Quinn RE (2015) High quality learning experiences. Available at: https://thepositiveorganization.wordpress.com/.../high-quality-learning-experiences/. Retrieved 5 July 2018
Suarman S (2015) Teaching quality and students' satisfaction: the intermediatory role of relationship between lecturers and students of the higher learning institutes. Mediterr J Soc Sci 6(2):626–632
Suarman S, Aziz Z, Yasin RM (2013) The quality of teaching and learning towards the satisfaction among the university students. Asian Soc Sci 9(12):252–260
Sultana RG, Watts AG, McGowan T (2016) The role of higher education in promoting national security, welfare, and prosperity. J High Educ Policy Manag 38(4):419-433. https://doi.org/10.1080/1360080X.2016.1181886
Szymenderski P, Yagudina L, Burenkova O (2015) The impact of an assurance system on the quality of teaching and learning—using the example of a university in Russia and one of the universities in Germany. High Educ Stud 5(5):15–25
Uganda Vice Chancellors' Forum (2016) Enrolment in higher education in Uganda
Wuni IY, Agyeman-Yeboah S, Boafo HK (2017) Poor facility management in the public schools of Ghana; recent empirical discoveries. J Sustain Develop Stud 11(1):1–30
Xaba MI (2012) A qualitative analysis of facilities maintenance: a school governance function in South Africa. South Afr J Educ 32:215–226
Zaman M (1998) Role of academic staff in improving teaching and learning. J Educ Dev 18(2):143–154
Zhang S, An N (2010) Analysis on the concepts and theories of the quality of graduate education. Asian Soc Sci 6(12):86–89

Springer Nature Customer Service Center GmbH
Europaplatz 3
69115 Heidelberg, Germany

The manufacturer's authorised representative in the EU is Springer Nature Customer Service Centre GmbH, Europaplatz 3, 69115 Heidelberg, Germany. If you have any concerns regarding our products, please contact ProductSafety@springernature.com

Printed and bound by CPI Group (UK) Ltd, Croydon, CR0 4YY
30/09/2026
02225598-0005